T0219076

CANCER STEM CELLS

Cancer Stem Cells

PHILOSOPHY AND THERAPIES

Lucie Laplane

 Harvard University Press   Cambridge, Massachusetts · London, England   2016

First printing

Library of Congress Cataloging-in-Publication Data
Names: Laplane, Lucie, 1984– author.
Title: Cancer stem cells : philosophy and therapies / Lucie Laplane.
Description: Cambridge, Massachusetts : Harvard University Press,
    2016. | Includes bibliographical references and index.
Identifiers: LCCN 2015042054 | ISBN 9780674088740 (alk. paper)
Subjects: LCSH: Cancer cells. | Stem cells. | Cancer—Treatment—
Philosophy. | Oncology.
Classification: LCC RC269.7 .L37 2016 | DDC 616.99/407—dc23
LC record available at http://lccn.loc.gov/2015042054

To my family

CONTENTS

Abbreviations   ix

Introduction   1

I. Cancer Stem Cells: A New Theory of Cancer

1 Cancer Stem Cells' Triumph   13

2 The CSC Theory   27

II. The Historical Emergence of the CSC Concept and Its Driving Role in Cancers

3 Teratocarcinomas and Embryonic Stem Cells   47

4 Leukemic and Hematopoietic Stem Cells   70

III. Debates on CSCs and Stem Cells: What Are They?

5 Origin, Stemness, and Stem Cells: The Meaning of Words   91

6 Stem Cell Identity   106

IV. The Identity of Stemness and Its Consequences for Cancer Therapies

7 If Stemness Is a Categorical or a Dispositional Property,
How Can We Cure Cancers?   139

8 If Stemness Is a Relational or a Systemic Property,
How Can We Cure Cancers?   156

Conclusion   179

Notes   191
References   197
Acknowledgments   243
Index   245

ABBREVIATIONS

AACR	American Association for Cancer Research
AML	acute myeloid leukemia
AML-CFU	acute myeloid leukemia–colony-forming unit
Ang-1	angiopoietin-1
APL	acute promyelocytic leukemia
ARC	Association pour la recherche sur le cancer
ATRA	all-*trans*-retinoic acid
BMP	bone morphogenetic protein
CFU-S	colony-forming unit–spleen
CLL	chronic lymphocytic leukemia
CML	chronic myeloid leukemia
CSC	cancer stem cell
dpp protein	decapentaplegic protein
EC	embryonal carcinoma
EMT	epithelial-mesenchymal transition
ES cell	embryonic stem cell
FACS	fluorescence-activated cell sorting
G-6-PD	glucose-6-phosphate-dehydrogenase
G-CSF	granulocyte–colony-stimulating factor
HGF	hepatocyte growth factor
HIF-1α	hypoxia-inducible factor-1 alpha
HPC	homeostatic property cluster

HSC	hematopoietic stem cell
IL-3R	interleukin-3 receptor
INCa	Institut National du Cancer
INFα	interferon alpha
iPS cell	induced pluripotent stem cell
ITGA6	integrin-alpha-6
NFκβ	nuclear factor kappa beta
NKT	Natural killer T
NOD-SCID	nonobese diabetic–severe combined immunodeficiency
PGK	phosphoglycerate kinase
PML	promyelocytic leukemia protein
ROS	reactive oxygen species
SAZ	segment addition zone
SCID	severe combined immunodeficiency
SDF-1	stromal-derived growth factor-1
STAP	stimulus-triggered acquisition of pluripotency
TGF-β	transforming growth factor–beta
THPO	thrombopoietin
TNFα	Tumor necrosis factor–alpha
Upd	unpaired
VEGF	vascular endothelial growth factor
WMISH	whole-mount *in situ* hybridization

CANCER STEM CELLS

INTRODUCTION

Cancer is a pandemic disease. The 2014 *Cancer Facts and Figures* of the American Cancer Society reports, "In 2014, about 585,720 Americans are expected to die of cancer, almost 1,600 people per day. Cancer is the second most common cause of death in the United States, exceeded only by heart disease, accounting for nearly 1 of every 4 deaths" (American Cancer Society 2014, 2). According to figures from the French association La Ligue contre le Cancer, one in two men and one in three women will be diagnosed with cancer before age 85. Of these, only 60 percent will recover.[1] These alarming data indicate the importance of the search for new cancer treatments. This issue is not new. In 1971, US President Richard Nixon declared a "war on cancer."[2] Progress achieved since then is undeniable, yet cancer mortality has declined by only 5 percent, while mortality due to cardiac conditions has declined by 64 percent, and flu and pneumonia by 58 percent (Kolata 2009). The cancer statistics of the National Cancer Institute show huge variation in the progress of five-year cancer survival statistics from 1975 to 2006. Some cancers, like prostate cancers or leukemias, show great increases in the five-year survival rate: from 66 percent in 1975 to 99.6 percent in 2006 for prostate cancers and from 33.1 percent to 60.8 percent for leukemias. But some cancers show little progress in our ability to cure them. Lung and bronchial cancers' five-year relative survival only increased from 11.4 percent to 17.5 percent, and pancreatic cancers from 3 percent to 7.3 percent, during the same timeframe. Others show no progress at all, such as endometrial cancers, where the five-year relative survival even decreased from 87.8 percent to 84.7 percent.

This critical context underscores the need for the development of new therapeutic strategies, among which are immunotherapies, anti-angiogenic therapies, and various targeting therapies. The latter category includes the "cancer stem cells" (CSCs) targeting strategy, the subject of this book. CSCs have been touted as the "new target in the war against cancer" (Lenz 2008). They "are cancer's 'Achilles Heel,'" according to Dr. Cordon-Cardo, a famous oncologist at Mount Sinai School of Medicine.[3] Given the attention that CSCs have received for their therapeutic potential in the war on cancer, we must ask the following questions: could CSCs represent a "cornerstone of a forthcoming therapeutic breakthrough?" (Sourisseau et al. 2014, 7; see also Chargari et al. 2012; Moncharmont et al. 2012) and, if so, what about these cells facilitate such a breakthrough?

CANCER STEM CELLS: EMERGENCE OF A
NEW THERAPEUTIC STRATEGY

What is a "cancer stem cell" and how could "targeting" them allow us to break the stalemate in the war against cancer (Haber, Gray, and Baselga 2011, 19)? Cancer stem cells are, as their name suggests, cells that combine two identities: they are both cancer cells and stem cells. Like other cancer cells, they carry various alterations, genetic or not, that make them dysfunctional. Like stem cells, they are able to self-renew and to differentiate, and they represent only a small fraction of cells. This raises a question regarding the CSC-targeting therapeutic strategy: how can the elimination of a small fraction of cancer cells cure a cancer?

To understand this potentially revolutionary therapeutic strategy, one must consider the theory that accompanies it: the "cancer stem cell theory." According to this theory, formulated between 1990 and 2000, only CSCs are capable of causing cancers. They do not constitute the whole of a cancer; rather, they represent a very small subset of cancers cells, but only *they* can induce its formation. The property of stemness explains this functional disparity: only cells capable of self-renewal and differentiation can initiate, develop, and spread cancers. Conventional therapies (chemotherapy and radiotherapy), whose goal is to eliminate as many cancer cells as possible, fail to effectively kill these very specific cells. Therefore, proponents of the CSC theory claim that therapeutic strategies against cancers should be reassessed and that we should instead target the CSCs.

More precisely, given that non-CSCs are devoid of any tumor-forming ability (tumorigenicity), the CSC theory conveys the idea that the elimination of all CSCs of a given cancer is "necessary and sufficient" to cure the patient (Reya et al. 2001, 110). This book examines the CSC theory and the therapeutic strategy it conveys. More specifically, it elucidates how targeting CSCs may or may not be "sufficient" to cure cancers.

Whether it is "necessary" to kill the CSCs is beyond the scope of this research. If it is true that cancers are initiated, developed, and maintained by CSCs only, then the elimination of CSCs is logically necessary to guarantee the definitive recovery of patients. Thus, asking whether the elimination of CSCs is "necessary" is asking whether the CSC theory is true. Only experiments, currently being conducted by biologists, will allow us to know whether the CSC theory is true and to which cancers it applies.

Thus, the question that this book addresses is precisely the following: if the CSC theory is true, then is it sufficient to target the CSCs in order to cure cancers? By unpacking and exploring the concepts of stem cell and CSC, my philosophical analysis exposes important distinctions in the way they are conceived. Evaluating the potential efficiency of the CSC-targeting strategy, as well as other therapeutic strategies such as niche targeting or differentiation therapies, relies on deciphering these distinctions and highlights the importance of collaborations between philosophers and scientists.

WHAT IS A STEM CELL?

What is the relationship between the identity of stem cells and the treatment of cancers? How can philosophical considerations shed light on the strategy of targeting CSCs?

Stem cells and CSCs raise philosophical questions regarding their identity because we still do not know exactly what they are. This might sound surprising since stem cells are touted as promising therapeutic tools and are already part of the current therapeutic arsenal for a number of diseases. Stem cell therapies attempt to use stem cells to treat diseases such as Parkinson's disease or Alzheimer's, as well as conditions such as brain or spinal cord injuries, blindness, and vision impairment. These types of reparative interventions using stem cells have raised enormous hope for combating degenerative diseases and injuries. However, except for bone marrow transplantations (transplantations of blood stem cells routinely used in the treatment of blood cancers), these therapies still have shown very little success in humans. To improve the use of stem cells as therapeutic tools throughout a wider range of disease and conditions, we critically need a better understanding of what stem cells are and how they work, including the nature of their stemness (i.e., their identity).

To understand why the identity of stem cells is problematic, it is first important to stress the heterogeneity of stem cells. Ernst Haeckel first framed the concept of stem cell *(Stammzelle)* in 1877 to describe the fertilized egg cell as the cell from which arise all the other cells of the developing organism (Maehle 2011). The use of the term "stem cell" rapidly evolved to refer to many other cells, both in the embryo and in the adult organism. The concept of stem cell now includes a wide spectrum of cells, normal or pathological (cancer stem cells), natural or artificial. Stem cell types contained in an organism may vary depending on the developmental period and on the tissue. In addition, all species do not have the same types of stem cells at the same developmental period.[4] Mammals are the taxon of interest for the issue of cancer treatments—this includes humans and mice (the main model organism for the study of CSCs). In mammals, three periods of development are usually distinguished, during which the body contains three different types of stem cells: totipotent, pluripotent and multipotent (see Figure I.1).

During the first cell divisions that follow fertilization (until the eight-cell stage), the cells of the embryo, called "blastomeres," are considered "totipotent stem cells" (Figure I.1A). Each of these stem cells is capable of giving rise to a complete organism.

Beginning at the 16-cell stage, two cell populations start to differentiate: the inner cells and the outer cells. At the 32-cell stage, a cavity appears within the embryo. As the cavity expands and the inner and outer tissues differentiate, the embryo enters the blastocyst stage (Figure I.1B), wherein trophoblast cells that form the outer wall (white cells in Figure I.1B) are distinguished from the "inner cell mass" (gray cells in Figure I.1B). When cells of the inner cell mass are extracted and cultured, they give rise to what are called "embryonic stem cells" or "ES cells" (Figure I.1C). These are "pluripotent stem cells" and can give rise to all the cell types of the adult organism. However, unlike totipotent stem cells, they cannot produce an entire organism because they are unable to develop the extraembryonic tissues, such as the placenta and the yolk sac, which are necessary for the development of the embryo. These extraembryonic tissues develop from the trophoblast cells. *In vivo,* the pluripotency of embryonic stem cells is extremely ephemeral—that is, their capacity to differentiate into all tissues of the organism is a transient property that is quickly limited. The cells of the inner cell mass give rise to the three germ layers (ectoderm, mesoderm, and endoderm) from which develop all the tissues of the organism (Figure I.1D). *In vivo,* pluripotent stem cells can persist throughout development and wreak havoc in the adult organism—through the formation of pathologies, such as teratocarcinomas. These pluripotent stem cells are the cancer stem cells of teratocarcinomas, from which it is possible to cultivate "embryonic carcinoma cell" lines (also referred to as pluripotent stem cells). More recently, pluripotent stem cell lines have been obtained from differentiated cells. These are the "induced pluripotent stem cells" or iPS cells (Figure I.1E).

Finally, following the formation of the three germ layers, the differentiation potential of stem cells is restricted to only certain tissues.

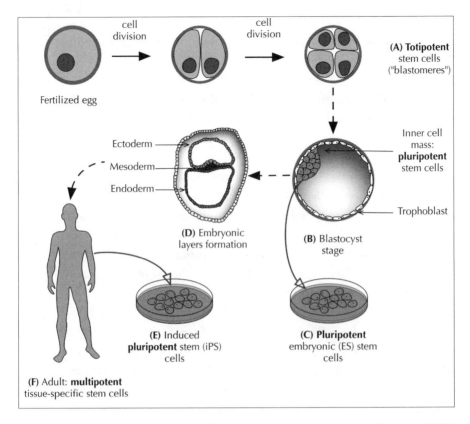

Figure I.1. Schematic representation of the different types of stem cells in humans. (Illustration © 2016 by Lucie Laplane.)

Thus, stem cells of the adult organism are commonly called "multipotent stem cells" or "tissue-specific stem cells" (Figure I.1F). Hematopoietic stem cells, for example, can give rise to all the different types of blood cells. These adult stem cells also have their cancerous counterparts: there are different cancer stem cells for different kinds of cancers, just as there are different stem cells for different kinds of tissues.

Given the heterogeneity of stem cells, several questions arise. The first one is a matter of definition: what do these cells have in common? Stemness is the common feature of all the stem cells. Stemness refers to two properties: the ability to self-renew, that is, the ability to produce at least one new stem cell during cell division, and the ability to differentiate, that is, the ability to produce more specialized cells. The question of whether stemness distinguishes stem cells from non-stem cells is an issue that must be clarified. Indeed, stem cells have very heterogeneous self-renewal and differentiation abilities compared with each other, and they are not the only cells to be able to self-renew and/or differentiate. Thus, a major question stem cell biologists struggle with is how to define stem cells.

The heterogeneity of stem cells and the nonspecificity of self-renewal and differentiation abilities raise another kind of question: do stem cells belong to a distinct natural kind, or are they a convenient category? This question is partially independent from the previous definition question because definitions do not necessarily tell us much about the identity of things. A definition expresses the properties shared by all the objects included in the definition. For stem cells, this common property is stemness. But the possession of common properties does not, for example, ensure membership within a natural kind. For example, the "fusiform" body shape is a common feature of dolphins, sharks, and ichthyosaurs, which are mammals, fish, and reptiles, respectively. Despite their distinct phylogenetic histories, these organisms are all tapered at the head and at the tail. Thus, possession of a fusiform body shape is a common property that can define a grade of organisms but could not be used to identify a natural kind. Therefore, the identity of a group of objects and its definition can be two separate issues. Additionally, objects have many kinds of properties. For example, "being fusiform" is a different kind of property than "being older than 60." Biological species are fusiform or not. In contrast, not all humans have the property "being older than 60," but they are all susceptible to enter, someday, in the group defined by the property "being older than 60." Thus, some properties, such as "being older than 60," can be acquired or lost. Others, like being fusiform, are more stable. What about stemness? Can a non-stem cell acquire stemness? Is stemness only dependent on what stem cells are, intrinsically, or does it involve relationships with other objects in the world, such as the cell's environment?

This book addresses all these questions to better understand stem cells and, more specifically, a particular type of stem cell: cancer stem cells. My philosophical investigation has at least two interests for cancer research and biologists. First, debates about stem cells can give the impression that there are as many conceptions of stem cells as there are stem cell biologists willing to engage those questions. Yet, by focusing on the identity of stemness, I arrive at a classification of four categories that subsumes all the different understandings of stem cells (categorical, dispositional, relational, or systemic). Second, this conceptual classification is highly relevant for cancer therapies as it allows predictions of the relative efficacy of diverse therapeutic strategies. To know which therapy to apply to a given cancer, one first needs to know which of the four stemness identities applies to it.

STRUCTURE OF THE BOOK

This book addresses the following question: if the cancer stem cell theory is true, then is it sufficient to target CSCs to cure cancers? The aim of this book is to show that the answer depends on the identity of stem cells or, more specifically, on the type of property stemness is. This aim is achieved through four parts.

The first part exhibits the strength of the CSC theory. Chapter 1 provides evidence of the spectacular development of this theory in the past 20 years, highlighting its favorable reception by the scientific community and its use by biotech and pharmaceutical companies to elaborate new therapeutic approaches. Chapter 2 exposes the structure of the CSC theory to explain its content—the major claims of the theory and the hypotheses on which they rely. It highlights the integrative and translational character of the theory that makes it so appealing: the CSC theory connects basic research to therapeutic applications, it might apply to all cancers, and it can explain various phenomena from a small number of hypotheses.

The second part deals with the historical emergence of the theory and the concepts of stem cell and CSC. Chapter 3 examines research over the past 150 years on the origin of solid cancers, in particular teratocarcinomas, which played a determinant role in the emergence of the concepts of cancer stem cell and embryonic stem cell. Chapter 4 discusses research since the 1950s on the origin of blood cancers and blood cells that led to the progressive construction of the concepts of leukemic stem cell and hematopoietic stem cell. This analysis demonstrates that the concepts of stem cell and cancer stem cell are historically ambiguous as to their meaning and their referent.

Following from the observation that the concepts of stem cell and cancer stem cell are historically ambiguous, the third part analyzes debates in the current scientific literature and shows that this ambiguity is still present. Chapter 5 reviews debates on the CSC concept, showing that it can refer to different things. Chapter 6 analyzes debates about the identity of stem cells, about their definition, and about whether they represent a natural kind. Together, they show that both concepts are subject to numerous investigations and debates related to their identity.

The fourth part of the book creates a classification of the different kinds of properties that stemness can be based on a philosophical analysis of the current scientific literature. According to this classification, stemness can be a categorical property, a dispositional property (Chapter 7), a relational property, or a systemic property (Chapter 8). This classification subsumes all the divergent views and data on stemness, giving a comprehensive account of stem cells and cancer stem cells that scientists can work with. Such a philosophical analysis of the concept of stemness is useful both to introduce clarity and precision into cancer stem cell research, as well as to distinguish and evaluate the potential efficacy of different therapeutic approaches. Different therapies will predictably have different efficacy depending on whether stemness is a categorical property, a dispositional property, a relational property, or a systemic property.

I

Cancer Stem Cells: A New Theory of Cancer

1

CANCER STEM CELLS' TRIUMPH

How important are cancer stem cells (CSCs) in cancer research? The concept of cancer stem cell has experienced a rapid development since the start of the twenty-first century. The existence of cancer stem cells was, at first, a debated hypothesis, particularly in solid cancers. In an interview, Peter Dirks, a pioneer in solid cancer stem cell research and a surgeon specializing in brain cancer at the Toronto Hospital for Sick Children, explained that in 1998, he had encountered some difficulties in financing his research when he decided to investigate the existence of CSCs in brain cancer. "'I had no funding,' he said. The granting agencies could not be swayed. 'It was donations from the families of my patients that kept the research going'" (Abraham 2006). In contrast with this early skepticism, this chapter shows that since the beginning of the twenty-first century, CSCs have experienced tremendous success as both a concept and a unit of inquiry, and are becoming an important topic in cancer research.

In the following sections, I provide evidence for the emergence of a research field focusing on CSCs along three independent lines of investigation:

1. A bibliometric investigation shows the yearly growth in the amount of publications on CSCs. These metrics indicate that from the year 2000 on, CSCs have garnered ever-increasing attention from scientific laboratories.
2. Evaluation of the American Association for Cancer Research (AACR) annual meetings, the most important international conventions in cancer studies, shows the emergence and growth of a research field focusing on CSCs. This emergence appears to be connected with the establishment of educational courses and resources.
3. Interviews conducted with CSC specialists show that funding has increased for CSC research. These data indicate that CSC research is becoming a competitive and well-funded field.

Bibliometrics

It is difficult to establish the exact moment at which the CSC concept emerged. It depends on what one aims to date: the first appearance of the term "cancer stem cell"? The first time biologists hypothesized that cancers originate and develop from stem cells? The first evidence for the existence and/or function of such cells?

Despite the difficulty of establishing an exact history, it is clear that the CSC concept has gained traction within the scientific community throughout the twenty-first century. Interest in this topic within the scientific community has increased exponentially since the early 2000s—beginning with a few publications per year in the early 1990s to more than 1,000 publications per year after 2005 (see Figure 1.1). A comparison with another emerging theory of cancers, namely the

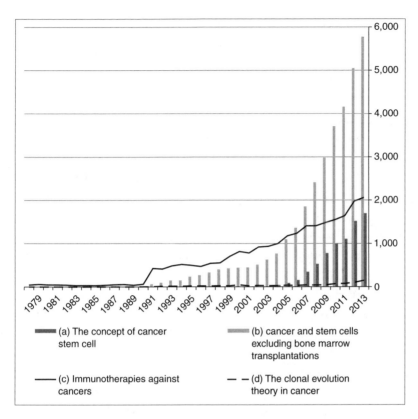

Figure 1.1. Number of publications by year of cancer stem cells compared to immunothera-pies and clonal evolution. This graph compares the outcomes of four searches done on *Web of Science* on June 30, 2014. All searches were done in the "Topic" research category, which includes titles, abstracts, and keywords of all the articles referenced in *Web of Science*. (a) Publications with the occurrence of the expression "cancer stem cell(s)." (b) Publications with the words "cancer" and "stem cell(s)," excluding publications with the words "bone marrow" and "transplantation(s)" that might use stem cells as a cure for rather than as the source of cancers. (c) Publications with the words "cancer" and "immunotherapy/immuno-therapies." (d) Publications with the words "cancer" and "clonal evolution." (Illustration © 2016 by Lucie Laplane.)

"clonal evolution theory" (see Chapter 2), and with promising therapies, namely "immunotherapies" which aim to stimulate the immune system, shows the high and rapid increase in interest for cancer stem cells.

The sheer volume of publications on CSCs provides solid evidence of its acceptance as a research domain, as well as a telling counterpoint to the skepticism that Dirks had encountered at his initial foray into the field in 1998. Nonetheless, a landslide of publications does not necessarily signify the mainstream acceptance or prominence of an idea. To find this type of information, we must look at the quality and prestige of the journals in which CSC research articles are published. CSC publications appear in all major journals, including *Nature* and *Science* (with 10 to 20 publications a year since 2003 in *Nature* and around 6 to 10 in *Science*), as well as in specialized journals (with, for example, more than four publications a year since 2007 in *Cell Stem Cell*).[1] Some articles on CSCs have been cited more than 3,500 times.[2] Additionally, between 2007 and 2014, at least 13 textbooks devoted to cancer stem cells and one series of 12 textbooks titled "stem cells and cancer stem cells" have been published in English (for the textbooks, see Wiestler, Haendler, and Mumberg 2007; Bapat 2008; Dittmar and Zänker 2008; Bagley and Teicher 2009; Farrar 2009; Majumder 2009; Yu 2009; Jordan 2010; Allan 2011; Scatena, Mordente, and Giardina 2011; Shostak 2011; Dittmar and Zänker 2013; Rajasekhar 2014; for the series, see Hayat 2012–2014).

These bibliometrics show that two important changes have occurred in CSC research since the millennium. First, CSC research has effectively emerged as a research domain, replete with an ever-expanding community of researchers. Research groups have popped up across the globe, creating a network of investigators from the Migrating Cancer Stem Cell Consortium created in 2007, a European network composed of researchers from the Netherlands, Italy, Germany, and Spain, to the Cancer Stem Cell Consortium established in 2008 (a gathering of Canadian and Californian researchers). In France, a network called the Cancer Stem Cell Network was established in 2006, and it originally gathered around 30 teams in Île-de-France. Likewise, at McMaster University in Hamilton, Canada, there is a Stem Cell and Cancer Research Institute, whose objective is to share the tools necessary for CSC research to cut the costs required to enter this research field.

The second change in CSC research trends that these bibliometrics show is that any objections that editors may have held against CSCs have been quelled. With publications on CSC research running in the magnitude of thousands per year, it is clear that both the research communities and the publishing communities upon which research thrives have reached a common agreement: CSCs constitute a real and important research domain.

Scientific Meetings and Education

Each year, the AACR gathers more than 16,000 cancer researchers at their annual meeting; this is the biggest yearly conference in cancer research. The continued increase of CSC-related talks and sessions at this gathering signifies the emergence and acceptance of the CSC research domain within the broader field of cancer research.

A comprehensive reading of the programs of the meetings that took place since 2004 shows that CSCs have been increasingly represented at the AACR annual meeting. CSCs began to pop up in 2004, starting as isolated presentations scattered throughout the program. No reference was made to the expression "cancer stem cell," but some sessions and posters dealt with stem cells and cancers. In 2006, the concept of CSC began to congeal into a coherent and autonomous research domain with the presence of a symposium (titled "Cancer Stem Cells") and a forum (titled "Identifying and Targeting Stem Cells"). In each subsequent year, until 2011, CSCs had gained more representation on the program, accumulating poster sessions, forums, "meet-the-expert" and educational sessions, and even an annual symposium (beginning in 2008 as a tribute to oncologist Bayard Clarkson: "The Bayard D. Clarkson Symposium on Stem Cells and Cancer").[3] Since 2011, CSC representation at the conference seems to have stabilized (see Table 1.1).

In addition to the broadening acceptance that CSCs have received within the cancer research community, they have also entered into the educational system, provoking an explosion of new textbooks and courses. This new field of research has now been established in the medical and biological education systems. Since 2007, the Life Sciences Learning Center at the University of Rochester has offered a class called "Stem Cells and Cancer" within its Cancer Education Project. The center provides a module for university teachers, complete with presentations and student activities and assignments. Likewise, the Crimson Foundation in Buenos Aires, Argentina, and the Cold Spring Harbor Laboratory offer courses and summer courses on CSCs.[4] In addition to the educational modules available online, researchers have created other learning tools to spread information about CSCs to biologists. For example, the journal *Nature Reviews Cancer* produced a poster titled "The Emerging Biology of Cancer Stem Cells," with CSC specialists as

Table 1.1.   Number and types of presentations on stem cells and cancers during the annual AACR meetings, 2004–2013.

	2004	2005	2006	2007	2008	2009	2010	2011	2012	2013
Symposia	1	1	2 (1)	1	1	2 (1)	1	3 (1)	4 (1)	1
Educational sessions	1		1	2	1 (1)	1	2 (1)	4 (4)	1 (1)	3 (1)
Meet-the-expert sessions		1	3	1	1 (1)	1 (1)		3 (1)	3 (2)	3 (3)
New concepts in organ site research		1		1			1 (1)	2		
Posters		2	2		5	6 (4)	7 (7)	7 (6)	7 (7)	8 (6)
Forums				1 (1)	1 (1)	2 (2)	1 (1)	1 (1)		1
Minisymposia					1 (1)	2 (1)	1 (1)	1	2 (1)	1 (1)
Methods workshops								1 (1)		2 (1)
Totals	2	5	8 (1)	6 (1)	10 (4)	14 (9)	13 (11)	22 (14)	17 (12)	19 (12)

Note: The numbers in parentheses enumerate those that mention the expression "cancer stem cell."

poster advisors (Hans Clevers, Connie Eaves, Richard Gilbertson, Gail Risbridger, and Jane Visvader). The poster is available on the websites of both *Nature Reviews Cancer* and Abcam, a UK-based antibody company that supported the production of the poster. Abcam provides hard copies for free upon request—pointing out a commercial interest in addition to the scientific interest that is spurring more research into CSCs (see below). And finally, the factsheet "Cancer: Disease of Stem Cells," produced by the partnership EuroStemCell, is available on their website in several languages and is designed to be accessible to a general audience.

A COMPETITIVE AND WELL-FUNDED FIELD

We have already seen that CSCs emerged during the early 2000s as a full-fledged research domain, replete with international collaborative networks, increasing numbers of high-stature publications, and support within the larger cancer research community. The only thing missing from this picture is finance—CSC research requires a massive amount of capital investment, and over the past few years, investors have risen to the challenge of building the financial infrastructure necessary to support this rapidly growing field. Billions of dollars have been invested in research on CSCs, with substantial capital directed toward basic research and exponentially higher funding poured into clinical trials.

The Cancer Stem Cell Consortium, a collaboration established in 2007 between CSC researchers in Canada and California, provides a good example of the extent of CSC research funding. Its strategic scientific plan, which is accessible through the consortium's website, states that it had a C$500 million budget for the period 2009–2014. In France, all the main institutions and associations financing cancer research (ARC, Cancéropôle, INCa, and Ligue Nationale contre le cancer) also finance CSC projects. In one of the largest funding decisions it ever made, Cancéropôle île-de-France teamed with INCa to support the French Cancer Stem Cell Network, with up to €2 million and €1.5 million, respectively.

In 2009, Dominique Bonnet, one of the pioneers of CSC research, explained,

For approximately five years, it is apparent that there is much more money for cancer stem cells research, this is obvious. . . . When I started my own laboratory (in England in 2001), the idea of cancer stem cell was not well developed yet, and there were more difficulties to find agencies interested in financing this kind of research.[5]

Bonnet's interpretation of a recent increased abundance of funding for CSC research is borne out in interviews with other current CSC researchers. Out of a cohort made up of 12 lab directors, researchers, post docs, and graduate students, only one person declared that she had difficulties in financing her research on CSCs. In this instance, she was denied because the funding body "thought that mentioning CSCs was only a strategy to get funding"; her following applications were successful.[6] Finding funding is no longer a difficulty that CSC researchers face; rather, the opposite—CSC research can be a lucrative endeavor. Several of the interviewees acknowledged that working on CSCs is now "a way to get money" for research.[7] The results of these interviews highlight that CSCs have become both a highly fundable and fashionable research domain.

THERAPEUTIC HOPES

While the establishment of a new research domain is an intriguing phenomenon, a striking aspect of the success of CSCs is the therapeutic potential that this research holds for the treatment of cancer. To establish the tight connection that the CSC theory has with a therapeutic potential, I will briefly discuss three phenomena:

1. A figure highly embedded in the current collective representation of CSCs, published in *Nature* in 2001, which outlines the treatment of tumors with traditional versus CSC-based approaches (Reya et al. 2001, 110, fig. 5)
2. The emergence of numerous public and private biotechnology societies that develop CSC-targeting drugs
3. The establishment of research programs for investigating CSCs within all major pharmaceutical companies

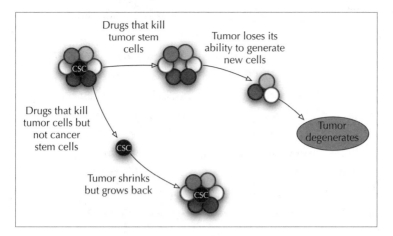

Figure 1.2. Cancer stem cells may be the key to an effective therapy. CSCs' escape from conventional therapies can result in relapses, whereas if they were killed, the tumor would lose its ability to maintain itself and would regress, leading to a cure. (T. Reya, S. J. Morrison, M. F. Clarke, and I. L. Weissman, "Stem cells, cancer, and cancer stem cells," *Nature* 414 (6859): 105–111, Figure 5. © 2001 Macmillan Magazines Ltd. Redrawn and reprinted by permission from Nature Publishing Group, Macmillan Publishers Ltd.)

A New Therapeutic Strategy

In 2001, Reya and colleagues published a review article in *Nature,* titled "Stem Cells, Cancer, and Cancer Stem Cells." In this article, the authors introduced a figure (see Figure 1.2), which has become a classic explanatory model for the CSC hypothesis (Reya et al. 2001, 110, fig. 5). The authors claim, through this figure, that CSCs could be the key to an effective therapy for targeting tumors, especially tumors that are prone to regrowth under traditional therapies (see Chapter 2).

The key idea behind this figure is the heart of CSC research—tumor growth (and nonresponse to traditional therapies) is a function of the presence of CSCs within the tumor. CSCs are not targeted by traditional therapies and can persist as malignant entities within the body, provoking cancer relapse. These traditional treatments, then, fail exactly because they do not take into account CSCs—the very objects that cause the cancer in the first place. Thus, if the CSC theory is correct, to avoid relapses, it would be necessary to change the therapeutic strategy and specifically target CSCs. Under this theoretical framework, by effectively eliminating CSCs, physicians would be able to ensure patients' long-term recovery, as the remaining cancer cells would naturally regress.

This figure is highly embedded in the current collective representation of CSCs and can be found within many subsequent CSC publications, on biologists' societies' web pages, in blogs, in open encyclopedias such as Wikipedia, and in pharmaceutical companies' and startups' marketing images. Moreover, the model presented in this image has been spread internationally and translated into several different languages.[8]

Therefore, the concept of cancer stem cell is tightly linked, in collective representation, to a new therapeutic strategy proposal (consisting of specifically eliminating CSCs), which gives hope of recovering from cancer.

Biotechnology Startups Take Their Chances

Biotechnology startup companies have massively invested in CSC research. They aim to develop drugs to target CSCs. Among these companies, the best known is OncoMed, which was launched in 2004 by some of the same people who published the figure highlighted in the previous section (Michael Clarke and Sean Morrison).[9] As of today, OncoMed has five drugs in clinical trials. Demcizumab (OMP-21M18), one of the first anti-CSC products developed by OncoMed, and tarextumab (OMP-59R5) are the most advanced in their clinical trials. Vantictumab (OMP-18R5), ipafricept (OMP-54F28), and anti-Notch1 (OMP-52M51) are not far behind. All of these target signaling pathways are known for playing important roles in stem cells and cancer stem cells— namely, Notch and Wnt pathways.

In addition to OncoMed, several other startups are attacking CSCs. Boston Biomedical (founded in 2006) has two drugs in clinical trials: one targeted at colorectal and gastric cancers, and one targeted at advanced malignancies. Boston Biomedical describes both of them as molecules that "block cancer stem cell (CSC) self-renewal and induces cell death in CSCs as well as non-stem cancer cells." Eclipse Therapeutics (founded in 2010) has developed a monoclonal antibody targeting CSCs formerly called ET101 (now BNC101). Bionomics acquired Eclipse Therapeutics for $10 million in 2012, and now BNC101 is about to enter clinical trials. Verastem (founded in 2010), a clinical-stage biopharmaceutical company "focused on discovering and developing drugs to treat cancer by the targeted killing of cancer stem cells," has three products in clinical trials: VS-6063, VS-4718, and VS-5584. The first two are inhibitors of a protein called focal adhesion kinase, whose expression is higher in CSCs than in normal tissue, while VS-5584 inhibits a signaling pathway, called PI3K/mTOR, which is described as "a key regulator in cancer progression and the survival of cancer stem cells."[10] Finally, Stemline Therapeutics, a company that claims that targeting CSCs "may represent a major advance in the fight against cancer,"[11] has two drugs in clinical trials. One, SL-401, is targeting the interleukin-3 receptor (IL-3R), a receptor overexpressed in leukemic immature cells and CSCs, and the other one, SL-701, is an immunotherapy that activates the immune system to attack tumor cells.

While these biotechnology companies were formed specifically to pursue CSC therapeutics, several other biotechnology companies, not specifically related to CSC therapeutic strategies, have also developed drugs for use against CSCs or have tested the effects of their drugs on CSCs. For example, Formula Pharmaceuticals, a company that specializes in oncology, has tested the effectiveness of its immunotherapy, FPI-01, on CSCs.

Moreover, KaloBios Pharmaceuticals, a personalized medicine company, developed drug KB004 to target CSCs. KaloBios Pharmaceuticals describes its new drug's potential: "what differentiates KB004 is its potential to attack tumors at their source by killing tumor stem cells, the tumor stromal cells that protect them, and the vasculature that feeds them. This unique activity provides the potential to generate durable responses by targeting the source of tumor cells."[12] Geron Corporation, a clinical stage biopharmaceutical company, developed a telomerase inhibitor called imetelstat (GRN163L) to treat blood malignancies. Imetelstat (GRN163L) entered clinical trials in 2005—the following year Geron announced that imetelstat also affects CSCs, and have recently demonstrated its ability to target CSCs in multiple myeloma, as well as pancreatic and breast cancers (Brennan et al. 2010; Joseph et al. 2010). And finally, ImmunoCellular Therapeutics, a clinical-stage biotechnology company, is developing novel cancer vaccine immunotherapies (ICT-107, ICT-121, and ICT-140) that target CSCs in glioblastoma, an antigen commonly associated with CSCs (namely, CD133), and ovarian cancer, respectively.

Pharmaceutical Companies Address the Field of CSCs

Following biotechnology companies, massive pharmaceutical companies have entered the fray—devoting enormous resources to pursue this line of cancer therapeutics. Outlined here are just a few examples of the extensive efforts that these companies are putting into CSC therapeutic research.

Sanofi-Aventis started working with the Institute of Hematology of the Chinese Academy of Medical Sciences located in Tianjin, China, in 2007, to isolate leukemic stem cells in acute myeloid leukemia and to eventually develop antibodies that specifically target CSCs.

In 2008, GlaxoSmithKline signed a $1.4 billion agreement with OncoMed Pharmaceuticals to develop antibodies that could specifically target CSCs. This agreement had been described as "the largest deal ever for a preclinical stage biotech company" (Madrigal 2008). At that time, OncoMed did not have yet any products that could be marketed soon (Solovitch 2008).

In March 2009, Novartis signed a research and development agreement with the UK biotechnology company EpiStem, which specializes in personalized medicine and epithelial stem cells, to develop screening tests for medicines that target CSCs. Under the terms of the agreement, Novartis paid Epistem an upfront cash payment of $4.0 million and provided research funding for at least two years.

In 2009, Pfizer Global Research and Development began collaboration with the Ontario Institute for Cancer Research and the Princess Margaret Hospital for a project called POP-CURE (PMH-OICR-Pfizer-CURE). The project aims to identify molecular signatures that could help produce new targeting drugs and, in particular, CSCs-targeting drugs. Pfizer contributed $6 million. "This exciting collaboration will join the world class genomics and informatics programs at OICR, cutting edge research in cancer stem cell biology and functional genomics at OCI/PMH and the world's largest pharmaceutical company in a concerted effort to bring new therapies to colon cancer patients worldwide," said Dr. Ben Neel, who serves as a principal investigator with the POP-CURE project.[13] Since 2012, Pfizer has also collaborated with Verastem for the clinical assays of their VS-6063 product candidate.

Beginning in 2011, the Swiss pharmaceutical company Roche provided researchers at the University of California, Los Angeles with advanced technologies (including the latest generation microarray systems from Roche NimbleGen, high-throughput screening instruments, genetic expression profilers, and exome sequencing technologies) and bioinformatics support for their research projects on stem cells and cancer. This agreement is meant to provide Roche with new predictive biomarkers, which they can use as targets for future therapeutics and diagnostics.

In 2011, Merck developed a collaboration with a CSC research team headed by Andreas Trumpp at the German Cancer Research Center (DKFZ) in Heidelberg. Merck provides Trumpp with new active ingredients, which he then uses to try to kill off the activated CSCs.

In 2011, Dainippon Sumitomo Pharma signed a product option license deal with Boston Biomedical's product BBI608. Under this agreement, Dainippon Sumitomo Pharma receives license fees for all oncological prescriptions of BBI608 in Japan and exclusive right of negotiation for the compound for the United States and Canada. Under the terms of the deal, Boston Biomedical received $15 million upfront and clinical trial support on signing. In 2013, Dainippon Sumitomo Pharma struck a deal to buy Boston Biomedical for $200 million upfront plus milestone payments that could top $2.4 billion, for a total of $2.6 billion.

CONCLUSION

In this chapter, we have seen evidence that CSCs have been established as a prominent and thriving research domain. Publications in a variety of journals—both specialized (i.e., *Oncogenesis, Tumor Biology, Cancer Cell*, etc.) and more general (i.e., *Nature, Science, PNAS,* etc.)—have exploded in recent years, as has the presence of CSC research at the largest annual cancer research conference in the world (the AACR) and within educational institutions. Hundreds of millions of dollars have funded CSC research within startup biotechnology companies, and large pharmaceutical companies have invested billions of dollars to pursue CSC research. All of these advances in CSC research are possible because CSCs constitute the core of research on and development of new medicines and therapeutic techniques for cancers. The innovative strategy for targeting different cancers by attacking CSCs has provided new hope that a complete cure for cancers can be found or, more precisely, that one therapeutic strategy (killing the CSCs) can be applied to develop new curative treatments for many forms of cancers.

2

THE CSC THEORY

That CSCs represent a coherent and thriving research domain is clear—they are the subject of active inquiry for a growing international network of researchers and the locus of billions of dollars in funding. But what are CSCs exactly? How do biologists define them? And what role do they play in cancers that make them so important to study and target? In this chapter, I analyze the theoretical framework that biologists have built around CSCs, which I refer to as the CSC theory.[1] I will show that the CSC theory is a well-integrated and powerful theory that contains three models (a model of cancer development, a model of cancer relapse, and a model of therapy) that rely on a series of hypotheses about the nature of CSCs and how they function during carcinogenesis. This chapter highlights the strength of the CSC theory, which comes from its ability to explain various phenomena (cancer development and propagation, as well as cancer relapses) from a very limited number of hypotheses and from its ability to connect basic research to biomedical interventions by suggesting a new therapeutic strategy for cancers.

WHAT IS A CSC?

In 2006, the American Association for Cancer Research (AACR) organized the first international meeting focused on cancer stem cells. The aim of the meeting was to discuss the data suggesting that cancers develop from cancer stem cells, a small subset of cancer cells with stem cell properties, the so-called cancer stem cell model. This model of cancer development relies on the existence of cancer stem cells, which begs the following question: what are cancer stem cells? The meeting participants reached a consensus in giving a functional definition of CSCs: "[A cancer stem cell is] a cell within a tumor that possesses the capacity to self-renew and to cause the heterogeneous lineages of cancer cells that comprise the tumor" (Clarke et al. 2006, 9340). This definition is very close to one that Tannishtha Reya and colleagues proposed in their oft-cited *Nature* article from 2001. They defined CSCs as "rare cells with indefinite proliferative potential that drive the formation and growth of tumours" (Reya et al. 2001, 105). Reya and colleagues implicitly assumed that CSCs drive tumor formation and growth by producing the heterogeneous cell lineages that occur within tumors.

Both the AACR meeting members and Reya et al. (2001) define cancer stem cells through four major propositions:

a. They are capable of self-renewal, thus producing new CSCs.
b. They are capable of differentiation, thus producing cells of different phenotypes.
c. They represent a tiny subpopulation of cells, distinct from other cancer cell populations and are, in theory, isolatable.
d. They initiate cancers.

These four propositions, as defined by the AACR and Reya et al. (2001), have come to constitute the way in which CSC researchers define their work. While this definition has been fruitful for moving the field forward, the four propositions are laden with assumptions about both the nature of CSCs and their functions during carcinogenesis. To get a clear picture about what constitutes a CSC and how they can be fruitful for constructing a model of carcinogenesis, we need to parse these propositions.

Propositions (a) and (b) specifically attribute two properties (self-renewal and differentiation) to CSCs—i.e., within cancers, they are properties that are held only by CSCs and not by other cells. Together, these two properties form what biologists call "stemness," which is defined as follows:

If a cell is capable of self-renewal (a) and differentiation (b), then it is a stem cell.[2]

Thus, stating that CSCs have properties (a) and (b) is equivalent to saying that they have "stemness" (marked as "P_s" in the figures). In other words, the CSC theory of cancer relies on a hypothesis (H_1); namely, that a particular category of cells, called CSCs, exists and is defined by the properties of self-renewal and differentiation, that is to say, by "stemness."

While propositions (a) and (b) characterize CSCs without referring to other cells, proposition (c), according to which CSCs are a tiny subpopulation distinct from other cancer cells, is relational: it implies a comparison between CSCs and *other* cancer cells. Thus, proposition (c) distinguishes between two classes of objects: CSCs and cancer non-stem cells (marked as "non-CSCs" in the figures). This proposition is not directly implied by the previous propositions (a) or (b).

It may be the case that empirical observation will show that all cancer cells have properties (a) and (b), in which case "cancer stem cell" would become equivalent to "cancer cell." Thus, proposition (c) involves a hypothesis (H_2), which is adjacent to the hypothesis that CSCs exist (H_1). Hypothesis H_2 claims that most cancer cells are not CSCs. Thus, proposition (c) is not simply a statement about CSCs or other cancer cells; it is also a claim about cancers more generally: cancers are made up of at least two distinct classes of cells, cancer stem cells and cancer non-stem cells. Properties (a) and (b) enable us to distinguish between those two classes of cells. Properties (a) and (b) also allow us to distinguish CSCs from non-CSCs, provided that biologists find a reliable way to identify the cells that possess these properties (this issue is discussed in Chapter 6). Moreover, proposition (c) adds a quantitative feature by implying that CSCs represent a "tiny subpopulation" of all cancer cells. Thus, within any given cancer, there are more non-CSCs than CSCs. Therefore, proposition (c) implies a second hypothesis, according to which not all cancer cells are CSCs (H_2), which is coupled with the hypothesis that CSCs exist (H_1).

Finally, proposition (d), which holds that CSCs initiate cancers, involves both CSCs and cancers in general—CSCs initiate cancers, and cancers grow from CSCs. This proposition is considered a corollary for properties (a) and (b): CSCs are able to initiate tumors because they have self-renewal and differentiation abilities (see Figure 2.1). These are the properties by which cells grow tissues. Because of hypothesis H_2 (not all cancer cells are CSCs), proposition (d) implies CSCs are necessary for cancers to develop. In other words, proposition (d) not only claims that cancers *can* originate in CSCs, but it also claims that cancers *always necessarily* develop from CSCs and *never* from non-CSCs.

Taken together, propositions (a), (b), (c), and (d) show that CSCs, defined by properties (a) and (b), are involved in a theoretical framework characterized by properties (c) and (d). CSCs constitute the foundation of a model of carcinogenesis. I suggest that we should distinguish between the concept of CSC and the CSC model of carcinogenesis:

> *Concept of cancer stem cell: cancer stem cells are cancer cells capable of self-renewal (property (a)) and of differentiation (property (b)).*

> *CSC model of carcinogenesis: cancers are initiated and maintained by cancer stem cells.*

Distinguishing between the concept of CSC and the model of carcinogenesis enables us to highlight the difference, among the descriptions biologists have given so far, between the propositions that characterize what a cancer stem cell is and the propositions that use the CSCs to characterize the way cancers develop (see Figure 2.1). These distinctions are important because debates within the CSC research community do not always rely on the same propositions and/or hypotheses—a point that will be addressed in detail in the following chapters. Distinguishing between the concept of CSC and the models that use CSCs for

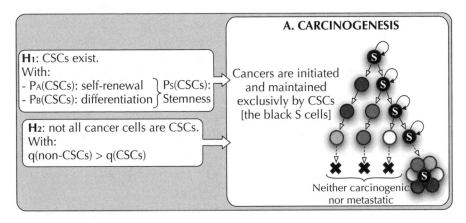

Figure 2.1. Structure and content of the CSC theory: the CSC model of carcinogenesis and the hypotheses on which it relies. (Illustration © 2016 by Lucie Laplane.)

carcinogenesis and therapeutics also makes it possible to highlight the conditions of validity for the hypothesis of the existence of CSCs and the ones specific to each model that relies on CSCs. For example, it is possible that CSCs exist (in this case, hypothesis H_1 would be correct), without the model based on CSCs being correct as well. Or CSCs could exist without driving cancer development. Distinguishing the conditions of validity is important to remind us that some propositions can be proven wrong without making the entire CSC theory false. It also gives biologists the ability to independently assess and discuss what CSCs are and the roles they may play in cancers.

A NEW MODEL OF CARCINOGENESIS

In the previous section, I highlighted that the characterization of CSCs both defines the CSCs as a kind of stem cell and involves them in a larger theoretical framework as the drivers of cancer development. Proponents of the CSC theory framed a model of carcinogenesis, according to which cancers develop from CSCs exclusively, and which relies on the hypothesis that cancers contains both CSCs and non-CSCs. Pursuing the analysis of the CSC theory, in this section, I explore the differences between the CSC model of carcinogenesis and the classical conception of cancer development. I show that the CSC model of carcinogenesis is more parsimonious than the classical conception of cancers, which strengthens the CSC theory. The CSC model of carcinogenesis unifies a number of explanations—low clonogenicity, heterogeneity of cancer cells, and non-metastatic disseminated cells—that require the use of many different models in the classical conception. Finally, I analyze the two other models contained in the CSC theory—namely, the CSC model of cancer relapse, according to which CSCs are responsible for relapses, and the CSC model of anti-cancer therapy, according to which it is necessary and sufficient to kill the CSCs in order to cure cancers. I will stress that these two models rely on at least one supplementary hypothesis— namely, that CSCs resist classical therapies.

The CSC Model of Carcinogenesis

In the 2001 *Nature* article mentioned above, Tannishtha Reya, Sean Morrison, Michael Clarke, and Irving Weissman schematized the difference between two models of carcinogenesis: the classical model and the CSC model (Figure 2.2). Reya et al.'s representation (Figure 2.2a) suggests that within the classical model of cancer, almost all cancer cells may contribute to the growth of the tumor to which they belong and can produce a new tumor (i.e., most cancer cells are capable of carcino-

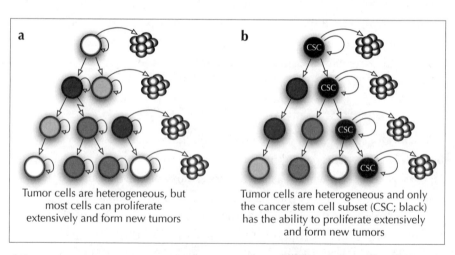

Tumor cells are heterogeneous, but most cells can proliferate extensively and form new tumors	Tumor cells are heterogeneous and only the cancer stem cell subset (CSC; black) has the ability to proliferate extensively and form new tumors

Figure 2.2. Schematic representation of the classical model (a) and the CSC model of carcinogenesis (b). (T. Reya, S. J. Morrison, M. F. Clarke, and I. L. Weissman, "Stem cells, cancer, and cancer stem cells," *Nature* 414 (6859): 105–111, Figure 4. © 2001 Macmillan Magazines Ltd. Redrawn and reprinted by permission from Nature Publishing Group, Macmillan Publishers Ltd.)

genesis). This graphic representation highlights two important hypotheses implicit within the classical model. First, that all cells are capable of self-renewal (this phenomenon is shown through circular arrows that begin and end with the same cell). Second, that different cell types are able to initiate new tumors (the different tones of gray among cells represent the heterogeneity among the cancer cells).

In contrast, the representation of the CSC model of carcinogenesis (Figure 2.2b) suggests that only CSCs are capable of self-renewal and that only CSCs are capable of generating new tumors.

Within this description of the two models of carcinogenesis, the causes required to start a new tumor are the same: the capability of self-renewal and differentiation. It is for this reason that the previous section stressed that the initiation of a cancer is considered a corollary to the property P_S (self-renewal and differentiation). The main difference between the classical model and the CSC model is the attribution of the property P_S to different cell types. The classical model attributes P_S in a stochastic way, or based on probability, to all cells and hence is usually referred to as a "stochastic model." The CSC model, on the other hand, attributes P_S to a sole particular cell type, the CSCs, and is usually referred to as a "hierarchical model," with CSCs at the apex of a hierarchy of immature to differentiated cell types, wherein all other cells are descendant from CSCs. Another difference between P_S in the classical model and in the CSC model is the mechanism through which different cells are produced; in the former, differentiation among cells happens through mutations, whereas in the latter, it is an intrinsic ability of CSCs.

The CSC Model Is More Parsimonious Than the Classical Model

The differences between the two models have significant consequences for our understanding of carcinogenesis. This section discusses the greater parsimony of the CSC model relative to the classical model, that is, the CSC model explains more independent classes of facts with less hypotheses. More precisely, both classical and CSC models make it possible to explain the weak clonogenicity of cancer cells, but only the CSC model provides direct explanations for heterogeneity among cells and the presence of disseminated but non-metastatic cancer cells.

(A) CLONOGENICITY. Clonogenicity refers to the proliferation ability of single cells—i.e., the ability of a cell to produce a colony of cells ("clones") that are descended from a single cell. Cancer cells have a low clonogenicity, that is, few cancer cells actually divide extensively and produce clones in clonogenic assays (see Bruce and Van Der Gaag 1963; Bergsagel and Valeriote 1968; Park, Bergsagel, and McCulloch 1971). The explanation of weak clonogenicity drawn from the classical model relies on a weak probability of expressing the property P_s. In other words, all the cancer cells are, in principle, able to form a clone, but the probability that they will do so is very low. In contrast, the CSC model explains low clonogenicity by the small number of cells that have property P_s (i.e., CSCs). That is, cancer cells lack the ability to extensively proliferate, except CSCs, which represent a tiny subset of cancer cells and are able to long-term self-renew and therefore to produce clones. Data in favor of this explanation of low clonogenicity in cancers are presented in Chapter 4 and discussed in Chapter 5.

While both models can account for clonogenicity, there are two challenges to our understanding of cancer that only the CSC model can explain without referring to ad hoc hypotheses: first, the heterogeneity of the cells that compose tumors and, second, the presence of non-metastatic cancer cells found at a distance from the primary tumor's site.

(B) INTRATUMOR HETEROGENEITY. Tumors are composed of heterogeneous cell lineages—they contain within them cells with many different phenotypes, degrees of motility, and levels of gene expression and metabolism. In the CSC model, the existence of heterogeneous cells in a cancer is a predictable consequence of the CSCs' ability to differentiate.

The heterogeneity of cancer cells poses a significant challenge for the efficacy of cancer treatments because of the difficulty of designing a therapeutic strategy that can equally affect all types of cells.

In contrast, the classical model does not directly provide an explanation for cellular heterogeneity. Instead, in the classical conception of cancers, the explanation for cellular heterogeneity relies on the integration of an independent model—the "clonal evolution model," according to which cancer cells evolve as a consequence of mutations and genetic and genomic instability. This model, also discussed in Chapters 3 and 5, was framed independently by Peter Nowell (1976) and John Cairns (1975). Based on the hypothesis that cancer cells that mutate play an important role in cancers, as well as on the observation that cell karyotypes are heterogeneous in cancers, Nowell and Cairns hypothesized that cancer cells evolve like species, via mutations that lead to the emergence of new heterogeneous clones. While the clonal evolution model is not incompatible with the CSC theory of cancer, its necessity in order for the classical model to explain cellular heterogeneity makes this model less parsimonious.

(c) Non-metastatic disseminated cancer cells. Since the 1950s, researchers have shown that "cancer cells exfoliate into body fluids, including blood and lymph, in a very high percentage of patients," but that "not all disseminated cells develop into metastases and . . . in many patients none produce metastatic lesions" (Southam and Brunschwig 1961, 971; see also Salsbury 1975). These shed cancer cells are called non-metastatic. Reya et al. (2001) highlighted that such non-metastatic cells, which are a bit of a mystery for the classical model, are logically explained by the CSC model.

The CSC model can explain disseminated non-metastatic cancer cells by reference to propositions (d) and (c)—that is, that only CSCs can initiate the development of a cancer and that CSCs are rare. If those two propositions are true, then "only the dissemination of rare cancer stem cells can lead to metastatic disease" (Reya et al. 2001, 109).

In contrast, according to the classical model, each cancer cell is capable of starting a cancer. They are thus expected to do so, sooner or later. Therefore, to explain the low prevalence of metastasis in cancer patients with exfoliated cells, biologists had to assume an additional hypothesis. This hypothesis was based on the ability of the immune system to target cancer cells in low numbers—i.e., the immune system should be very effective when it comes to targeting and destroying disseminated cancer cells before they produce a detectable tumor (Hanahan and Weinberg 2000; Reya et al. 2001, 110). While seemingly plausible, this hypothesis conflicts with an explanation of cancer relapse, as highlighted in the following section.

In summary, the CSC model of carcinogenesis provides a simpler explanation than the classical model for several phenomena of cancer—in particular, carcinogenesis, the weak clonogenicity of cancer cells, the heterogeneity of cells within tumors, and the presence of non-metastatic cancer cells that have disseminated from a primary tumor. In this respect, the CSC model of carcinogenesis is more parsimonious than the classical model of carcinogenesis, in that it can explain more phenomena with fewer hypotheses.

A NEW MODEL OF RELAPSE ASSOCIATED WITH A NEW MODEL OF THERAPY

In the preceding sections, I discussed how the concept of CSC was used to build a model explaining carcinogenesis. As frequently written in reviews on CSCs, CSCs are "thought to be responsible for cancer initiation, progression, metastasis" (Chen, Huang, and Chen 2013, 732), as well as for "recurrence of all cancers" (Kasai et al. 2014, 2).[3] The possible role of CSCs in cancer recurrence, or relapse, has given rise to two coordinated CSC models, one of cancer relapses and one of effective anti-cancer therapies. To ensure their explanatory and predictive value, these two models, which are parts of the general CSC theory, require the introduction of additional hypotheses—either that CSCs come from mutated stem cells or that they are resistant to therapies. As a consequence, if these hypotheses are false, the models of relapse and therapy can be false, without affecting the truth-value of the CSC model of carcinogenesis.

Resistance to Therapies and Relapses

In their elaboration of the CSC theory, Reya et al. (2001) developed two models. One is the model of carcinogenesis discussed above (see Figure 2.2), and the other is a model of anti-cancer therapies (see Figure 1.2). This latter model is composed of two parts: one is the explanation of cancer relapses by CSCs, and the other one is a prediction of the efficiency of CSC-targeting therapeutic strategies. Their argument goes as follows:

> It seems that normal stem cells from various tissues tend to be more resistant to chemotherapeutics than mature cell types from the same tissues. . . . If the same were true of cancer stem cells, then one would predict that these cells would be more resistant to chemotherapeutics than tumour cells with limited proliferative potential. Even therapies

that cause complete regression of tumours might spare enough cancer stem cells to allow regrowth of the tumours. (Reya et al. 2001, 110)

The idea that "in order to cure cancer, it is necessary and sufficient to kill cancer stem cells" follows from Reya et al.'s explanation of relapses (2001, 110). Ten years later, the explanatory logic of relapses by proponents of the now well-known CSC theory was the same:

> CSCs are of clinical significance as it has been shown that they are more resistant to both chemotherapy and radiotherapy than other malignant cells (Elrick et al. 2005; Vlashi et al. 2009). This may be a biological feature retained from normal tissue stem cells that natively possess various strategies to evade chemotherapy.... If CSCs are capable of evading adjuvant treatment then disease recurrence is likely where even a few tumour-initiating cells remain after therapy. (Buczacki, Davies, and Winton 2011, 1254)

This explanation of relapse relies on two propositions:

a. Resistance to therapies: CSCs escape therapies.
b. Relapses: CSCs are responsible for the regrowth of the cancer (I will refer to this as the "relapse").

Proposition (a) requires the insertion of an additional hypothesis in the CSC theory, whereas it is a logical consequence of the clonal evolution theory in the classical conception of cancers. Meanwhile, proposition (b) is better explained by the CSC theory than by the classical view.

(A) RESISTANCE TO THERAPIES. Cancer cells can resist or escape therapies in two ways. They can resist therapeutic interventions in such numbers that the therapy is considered a failure. Or, they can escape therapies in few numbers, in which case resistance might happen without being noticed at the clinical level—this latter case is one possible explanation for cancer relapse after remission.

In the classical view, resistance of cancer cells to therapies is explained by the clonal evolution hypothesis. In the clonal evolution model, cells are subject to multiple mutations that can lead to the acquisition or loss of properties, some of which would allow cancer cells to better resist, or escape, therapies (e.g., Aguirre-Ghiso 2006; Baguley 2010; Berridge, Herst, and Tan 2010). Therefore, the clonal evolution model provides explanations for both the intracellular heterogeneity among cancer cells and resistance to therapies.

In the CSC theory, the explanation of resistance relies on an analogy between stem cells and CSCs. Stem cells have been shown to resist therapies (Wicha, Liu, and Dontu 2006), and so, this analogy relies on the assumption that resistance is conferred to stem cells by the properties of stemness. If stemness is not the explanation for why stem cells are able to resist cancer therapies, then this analogy may be misleading.

Many cellular properties and processes participate in resistance to therapies, such as overexpression of anti-apoptotic proteins (Liu et al. 2006; Hambardzumyan et al. 2008), quiescence (Graham et al. 2002; Holtz, Forman, and Bhatia 2005; Copland et al. 2006), improved DNA damage response (Bao et al. 2006; Zhang, Atkinson, and Rosen 2010), higher capacity for drug efflux (Hirschmann-Jax et al. 2004; Huss et al. 2005; Abbott 2006), localization in a hypoxic niche (Calabrese et al. 2007), and lower concentration of reactive oxygen species (ROS) that mediate cell destruction by radiation (Diehn et al. 2009; for a review, see Vidal et al. 2014). Researchers have observed all of these properties in normal and cancer stem cells. The nature of the relationships between self-renewal and differentiation properties (stemness) and those cellular properties related to resistance remains unknown. At least three different types of relationships could exist between the properties witnessed in therapy-resistant cells and stemness:

1. Causal connection: these properties can be causally connected, such that properties that confer resistance to therapies would come with self-renewal and differentiation properties. In this case, because CSCs have the property of stemness, they necessarily have at least some of the resistance properties.
2. Contingency: these properties could be merely contingent. For example, cancer cells are able to survive in environments that lack oxygen, called hypoxic niches, regardless of stemness. If they can survive in hypoxic niches, they can escape therapies. Such an escape has nothing to do with stemness.

3. Inheritance: if CSCs are normal stem cells that become cancer cells due to mutations (a hypothesis extensively discussed in Chapter 5), then the mechanisms involved in resistance can be clonally inherited independently from self-renewal and differentiation properties.

One can only argue that the CSC theory makes it possible to explain CSC's resistance if the first relationship holds—that is, if there is a causal relationship between CSC's resistance properties and self-renewal and differentiation properties (stemness). If either relationship 2 or 3 holds, the explanation of CSC's resistance to therapies requires the addition of supplementary hypotheses to the CSC theory:

1. If the second relationship is correct—i.e., the properties are contingent—to explain relapses, one would have to postulate that CSCs initiate and maintain cancer (this is already part of the CSC carcinogenesis model) and that CSCs can resist therapies. CSC resistance would thus constitute a supplementary hypothesis and require explanation.

2. If the third relationship were correct—i.e., the properties of therapy resistance are acquired through mutation and transmitted throughout a cell lineage—one would need to hypothesize that CSCs initiate and maintain cancer (which the CSC model states) and that CSCs derive from normal stem cells that display resistance to therapy properties. The hypothesis that CSCs are derived from normal stem cells is not a new proposal—the first lab that provided evidence for the existence of CSCs (Canadian John Dick's team) claimed that the CSCs were derived from mutated hematopoietic stem cells (Bonnet and Dick 1997; see also Chapter 4). The 2001 *Nature* article from Reya et al. also explained cancer relapse by relying on this hypothesis. However, this hypothesis is a matter of intense debate.

Thus, without a demonstration of the relationship between stemness and the ability to escape therapies, the CSC theory requires a new hypothesis to provide an explanation for CSC resistance to therapies and thus to explain relapses:

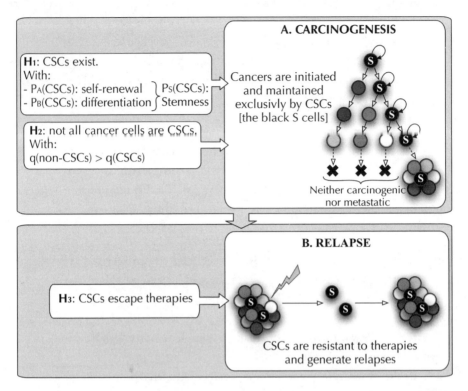

A. CARCINOGENESIS

H₁: CSCs exist.
With:
- P_A(CSCs): self-renewal ⎫ P_S(CSCs):
- P_B(CSCs): differentiation ⎬ Stemness

Cancers are initiated and maintained exclusivly by CSCs [the black S cells]

H₂: not all cancer cells are CSCs,
With:
q(non-CSCs) > q(CSCs)

Neither carcinogenic nor metastatic

B. RELAPSE

H₃: CSCs escape therapies

CSCs are resistant to therapies and generate relapses

Figure 2.3. Structure and content of the CSC theory: the CSC model of carcinogenesis, the CSC model of relapse, and the hypotheses on which they rely. (Illustration © 2016 by Lucie Laplane.)

Hypothesis 3 (H₃): CSCs are capable of escaping therapies.

(B) RELAPSES. Recall that CSCs exist within tumors in numbers far lower than other types of cells (hypothesis 2 of the CSC theory; see Figure 2.3). Because of this quantitative difference between cancer cells, if CSCs can survive therapies while other cancer cells die, it could make posttherapeutic CSCs hard to detect. This difficulty of detection then fuels clinicians to diagnose "remission," when in reality carcinogenic CSCs may still be present within the body. As CSCs are responsible for the tumor's growth, they are therefore capable of reinitiating a tumor, leading to cancer relapse. Therefore, if CSCs escape therapies, relapse is predictable from the CSC model of carcinogenesis.

In contrast, under the classical model, the probability that a single cell or a tiny group of cancer cells is capable of provoking a relapse is very low. This low clonogenicity of cancer cells leads to two predictions in the classical conception. First, the probability of a relapse depends on the number of cancer cells that evade therapy. The fewer number of cancer cells that evade therapy, the less probable a relapse. Second, when remission occurs, relapse should be unlikely, as most of the cancer cells are supposed to be killed. Yet, relapses are far from rare. General data are lacking because cancer registries do not collect recurrence information, but cancer survivors are considered "at risk for recurrence." To give some examples, 40 percent of patients treated for colorectal cancers experience recurrence, and the rate of recurrence among bladder cancer patients ranges from 50 percent to 90 percent (American Cancer Society 2012).

The high rate of relapses after apparently successful therapies contradicts two predictions of the classical conception of cancers. First, it goes against the prediction that relapses should be rare. Second, it conflicts with the hypothesis used to explain non-metastatic disseminated cancer cells—i.e., that the immune system efficiently kills cancer cells when they are few in number. The high rates of relapses, in addition to being difficult to explain, weaken the immune system hypothesis and, as a consequence, the ability of the classical conception to explain relapses. Therefore, relapses, as far as they involve tumor regrowth, are better explained by the CSC theory than by the classical theory.

CSCs: Therapeutic Targets?

Through addressing the issue of cancer relapse, CSC scientists have had a chance to both critique classical cancer therapeutic strategies and evaluation criteria as well as propose a new therapeutic strategy based on targeting CSCs. The classical cancer treatment strategy relies on the elimination of the "tumor mass," the efficacy of which is evaluated, during phase II clinical trials, through a set of rules called RECIST (Response Evaluation Criteria in Solid Tumor). RECIST consists of testing the capacity of drugs to reduce the size of a tumor, without regard to the constitution of the tumor or the cells that are targeted (Therasse et al. 2000; Eisenhauer et al. 2009). CSC specialists have questioned the validity of this strategy because tumor reduction does not take into account the elimination of different cell types within the mass. Thus, even if, under traditional therapies, a tumor decreases in size, CSCs

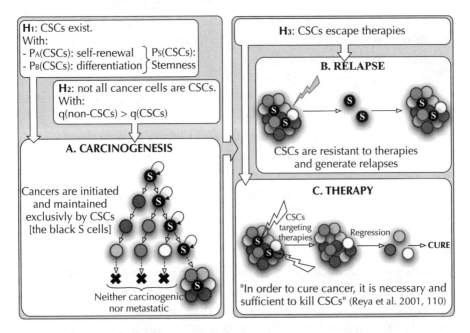

Figure 2.4. Structure and content of the CSC theory: the CSC theory, its models of carcinogenesis, relapse, and therapy, and the hypotheses on which those models rely. (Illustration © 2016 by Lucie Laplane.)

may not have been eliminated, which could lead to cancer relapse. In this light, the evaluative criteria set by RECIST do not seem like a reliable means of determining the effectiveness of treatments (Lenz 2008; Tu 2010; Vermeulen et al. 2012). Another problem that arises from this strategy is that by focusing on tumor reduction (i.e., remission), traditional therapies may be missing the true aim of cancer treatments—the complete elimination of cancer from the body. For this reason, CSC proponents have suggested a reevaluation of the goals of cancer therapies and a new model of cancer treatment along fundamentally different lines: "to cure cancer, it is necessary and sufficient to kill cancer stem cells" (Reya et al. 2001, 110).

The CSC therapeutic model (see Figure 2.4C) has one major advantage over previous models—if correct, the strategy that it dictates leads to patient *recovery*, not remission. Recovery is affected in this model by the targeting and elimination of CSCs, a strategy that would remove the possibility of relapse. Tumor degeneration, in this model, is thus a side effect of eliminating CSCs—once the source of carcinogenesis is removed, the consequence would be tumor degeneration.

Clinical trials of the drug imatinib offer an example of the difficulty of relapse within classical cancer treatment strategies and highlight how the CSC model favors recovery. Imatinib is a tyrosine kinase inhibitor specific to *BCR-ABL*, a mutation involved in the development of chronic myeloid leukemia (CML). Clinical trials with imatinib were highly successful, with 95 percent of the patients with late CML in the chronic phase achieving a complete hematological remission, 60 percent achieving a major cytogenetic response, and 41 percent achieving a complete cytogenetic response (Talpaz et al. 2013 and references therein). Unfortunately, approximately half the patients displayed a molecular relapse within six months following the discontinuation of imatinib (Cortes, O'Brien, and Kantarjian 2004; Rousselot et al. 2007; Stegmeier et al. 2010). Several studies have shown that imatinib was inefficient at targeting the CSCs of CML (Graham et al. 2002; Bhatia et al. 2003; Copland et al. 2006), leading biologists to argue that "a small population of cells, the so-called 'cancer stem cells,' are resistant to imatinib and therefore promote tumor relapse upon cessation of treatment" (Stegmeier et al. 2010, 546–547). Thus, the application of the CSC theory to therapeutic strategy could finally make it possible to design and develop therapies that lead to complete patient recovery.

CONCLUSION

The CSC theory comprises three models, relies on three hypotheses, and provides explanations and predictions for a range of phenomena. The most central model of the CSC theory is the CSC model of carcinogenesis (Figure 2.4A). This model, which states that cancers are initiated and maintained exclusively by CSCs, offers explanations of the development of cancers, of their heterogeneous cellular constitutions, of the low clonogenicity of cancer cell populations, and of the presence of non-metastatic disseminated cancer cells. This model relies on two hypotheses: that CSCs, defined by the property of stemness, exist (H_1) and that they represent only a tiny subpopulation of the cancer cells (H_2).

Based on these two hypotheses and the CSC model of carcinogenesis, biologists also framed two other models related to therapies. The

CSC model of relapses states that CSCs lead to relapse when therapies fail to kill them (Figure 2.4B). This model is partly deducible from the CSC model of carcinogenesis and its underlying hypotheses: relapses, like original cancers, necessarily develop from CSCs. However, to explain relapses after apparently successful therapies, the introduction of a complementary hypothesis (H_3) is necessary. H_3 states that CSCs can resist or escape therapies (because they come from stem cells, because they are stem cells, or for contingent reasons). Based on this explanation of the failure of cancer therapies, the CSC theory proponents have suggested a new model of therapy (Figure 2.4C), according to which a therapy can cure cancers if and only if it succeeds in killing the CSCs.

The CSC theory is a powerful theory of cancer because it provides a well-integrated and coherent set of explanations and predictions of the biology and treatment of cancers, requiring a low number of hypotheses. The next two chapters highlight the historical roots of the CSC theory and, in doing so, provide data to suggest that the hypotheses H_1 (CSCs, defined by stemness, exist) and H_2 (CSCs exist in low numbers in cancers) are true. These chapters also highlight ambiguities about the concepts of CSC and stemness and about which cells count as CSCs—topics that are addressed later in the book.

II

The Historical Emergence of the CSC Concept and Its Driving Role in Cancers

3

TERATOCARCINOMAS AND
EMBRYONIC STEM CELLS

Some proponents of the CSC theory trace its origin back to the "embryonic rest theory" in the nineteenth century, according to which cancers originate from embryonic cells that remained in the body throughout adulthood (the so-called embryonic rests). Biologists see theoretical continuity between the embryonic rest theory and the CSC theory because the embryonic cell remnants were characterized by a higher developmental potential than surrounding tissues and have thus been retrospectively conceived of as stem cells (e.g., Huntly and Gilliland 2005; Polyak and Hahn 2006; Tan et al. 2006; Wicha, Liu, and Dontu 2006; Gisselsson 2007; Eyler and Rich 2008; Sell 2009; Kreso and Dick 2014). Therefore, the embryonic rest theory and the CSC theory are two instances of a common hypothesis: cancers originate and develop from stem cells (or equivalent kinds of cells). This chapter highlights the great historical diversity in this general hypothesis of the stem cell origin of cancers.

I show that the CSC theory derives from at least three conflicting embryonic theories, namely, the "delocalized embryonic rest theory" of Robert Remak (1815-1865), the "connective tissue theory" of Rudolf Virchow (1821-1902), and the "superabundant embryonic rest theory" of Julius Cohnheim (1839-1884). These competing hypotheses of the stem cell origin of cancers persisted into the twentieth century through debates on the origin of teratocarcinomas wherein the "blastema theory" opposed the "germ cell theory." Research on teratocarcinomas by Leroy Stevens and Gordon Barry Pierce led to the first identification and characterization of pluripotent stem cells among cancer cells. Analysis of these different theories highlights the flexibility and permissiveness of the general hypothesis of the stem cell origin of cancers and raises a crucial question about the CSC theory: to what cells does the concept of CSC refer? Question that will be addressed in Part III of this book.

CELL THEORY AND ITS HYPOTHESES
OF CANCER'S ORIGIN

The embryonic rest theory emerged rapidly during the mid-nineteenth century as an explanation of cancer within the framework of the cell theory. Through a careful analysis of this emergence, I show that Robert Remak, Rudolf Virchow, and Julius Cohnheim's theories of cancer origin differ with regard to the cell that they describe as the cell of origin (delocalized embryonic cells, superabundant embryonic cells, or germ cells of the connective tissue) and/or with regard to whether they are general theories of cancer development or particular theories that explain the development of particular cancers.

The Emergence of Cell Theory

Two major and distinct principles, successively developed by different scholars, constitute the cell theory. The first principle describes the cell as the basic unit of living beings. While Matthias Schleiden (1804-1881) originally developed this principle for plants (Schleiden 1838), Theodor Schwann (1810–1882) extended its purview to include the animal kingdom (Schwann 1839). Schwann framed his hypothesis of cell formation through an analogy to crystals: the cytoblast (Schleiden used this name to refer to the nucleus) developed through crystallization of the cytoblastema (a liquid without structure, made up of sugar and mucus, which would be present between cells) before developing into a cell. Thus, to Schwann, cells arose from amorphous interstitial fluid (cytoblastema). The second principle, often stated through reference to the Latin phrase *omnis cellula e cellula,* claims that every cell comes from another cell. This principle arose from critiques that Robert Remak and Rudolf Virchow made about Schwann's hypothesis of the formation of cells from the cytoblastema.

Robert Remak, who worked on fertilized frog eggs—a material particularly well suited for observing cell division due to its size and transparency—criticized the role of the cytoblastema in cell development beginning in the 1840s. Through his observations, Remak gradually came to realize that cells originate by division from previous cells, a finding that removed completely the role that Schwann attributed to the cytoblastema in cell division: "everything happens as if, in the middle of the cell, there occurred a ligature, which broke the cells into two" (Remak 1855, cited in Rather 1978, 120–122). For complex reasons, including religious prejudice, Remak's observations of cell division went largely unnoticed (see Duchesneau 1987, 256–357). Therefore, although Remak had already framed the hypothesis that cells develop through division of preexisting cells, it was instead the critique of the esteemed pathologist, Rudolf Virchow, that led to the second principle of the cell theory in the mid-1850s.

Rudolf Virchow was a physician, whose main interest, in contrast to Schwann and Remak, was in pathology—i.e., the abnormal processes of development. Virchow considered diseases to be the effects of the alteration of normal physiological processes. His doctoral research on inflammation had led him to reject the prevailing theory of humors—a medical view in which the balance/imbalance of four bodily fluids governed health and disease. Virchow's rejection of humorism and willingness to construct a theory of disease based on deviation from physiological norms led him to criticize the theory of the emergence of cells from an amorphous cytoblastema (see Duchesneau 1987, in particular p. 295). This is how, in 1855, he came to claim the second principle of the cell theory: *omnis cellula e cellula,* a principle that applies to normal tissues as well as to neoplasias (Virchow 1855a, 1858).

Cell Theory and Cancers

The emergence of cell theory is intimately tied to the development of research on cancer, both conceptually and in terms of researchers. Julius Cohnheim, a noted pathologist and early cancer researcher, studied under Virchow at Berlin University, and both Virchow and Remak studied under Johannes Müller, whose assistant was Theodor Schwann.

Johannes Müller (1801–1858) introduced the use of the microscope to study cancers. In the nineteenth century, scientists sought to establish a rigorous empirical classification of neoplasias. Aware of his assistant Theodor Schwann's work on establishing cells as the basic unit of life, Johannes Müller aimed to found his classification on the observation of cells composing tumors (Müller 1838). His theory of the formation of neoplasias was based on humors and was closely related to Schwann's theory, to the extent to which it implies that cells develop through condensation of the cytoblastema from a *semen morbi*—the germs of a disease.

When Virchow proposed the second principle of the cell theory, his argument was aimed at the works of both Schwann and Müller, that is, against both the theory that cells develop from cytoblastema and against the humoral theory of pathologies. According to Virchow, tumors originate from the transformation of normal cells into tumor cells due to irritation or to an alteration of cell nutrition (see Duchesneau 1987, 343).

Remak, Virchow, and Cohnheim all agreed that the origin of tumors is dependent on the transformation of cells. However, they did not agree on which cells were the targets of transformation. To highlight the differences between these three authors, it is useful to begin with a controversy regarding a particular case of cancer. This controversy erupted in 1850 after Virchow diagnosed a tibial cancer as a primary epithelial tumor (Virchow 1850). Virchow's diagnosis was problematic because it challenged the prevailing theory of embryonic development at the time—germ layer theory. During development, embryonic cells differentiate into three layers through the process of gastrulation (see Figure I.1D). Remak called the external embryonic layer the "ectoderm," the intermediate embryonic layer the "mesoderm," and the internal embryonic layer the "endoderm."[1] Under germ layer theory, the cells and derivatives of each of these three layers are predestined and invariant: the ectoderm gives rise to the epidermis and to the nervous tissues; the mesoderm produces the connective tissues, muscles, and bones; and the endoderm produces the digestive and the respiratory tracts. Germ layer theory thus held it to be impossible that the epithelial cells that Virchow described came from transformed bone cells. How could they, when epithelium is an ectodermal tissue and bone is derived from the mesoderm? Therefore, several pathologists, such as Adolph Hannover (1814–1894), judged Rudolf Virchow's interpretation as "impossible": "a tissue could not be transformed into a different one" (Hannover 1852, 51; see also p. 21). Either the tumor was not epithelial, or it was not a primary tumor.

The Delocalized Embryonic Rest Theory (Robert Remak)

Remak agreed with Virchow on the diagnosis: the cancer cells Virchow had observed appeared to be epithelial cells, and the bone marrow seemed to be the primary site where the tumor had developed. This diagnosis constituted a great challenge to Remak, whose 1851 text had established the existence of the mesoderm and provided detailed observations about the derivatives of each of the germ layers (Remak 1855). To solve this patent contradiction, Remak suggested that this kind of tumor was derived from ectodermal embryonic cells that had become delocalized during the early stages of the embryo's development (Remak 1854, 172).

This explanation corresponds to the so-called embryonic rest theory, which I propose to call the "delocalized embryonic rest theory." This theory is similar to the CSC theory in two ways:

1. The differentiation potential of the cells that originated the cancer explains the characteristics of cancer cells.
2. The cells that originate cancers belong to a particular cell *type* (which we now refer to as stem cells).

Remak only defended this hypothesis in the few rare cases of tumors that looked like metaplasia, that is, tumors that contained cells coming from an embryonic layer different from the one that produced the tissue in which the tumor had developed. Thus, in contrast to the CSC theory, Remak's delocalized embryonic rest hypothesis is not a general theory of cancer development or a general theory of the origin of cancers—it is a highly specific and context-dependent hypothesis of how one type of cancer arises. Moreover, the conditions that forced Remak to adopt his hypothesis (i.e., the apparent contradictions with germ layer theory) do not play any active role in the construction or application of the CSC theory.

The Connective Tissue Theory (Rudolf Virchow)

While Remak was quick to reject Schwann's cytoblastema origins hypothesis of cells, Virchow accepted the doctrine until the end of the 1840s. And, it was not until 1854–1855 that Virchow began, as a result of his research on connective tissues, to develop the idea that cells derive from other cells. Connective tissues are a class of tissues that connect different types of tissues and organs. They are made up of cells, such as fibroblasts and adipocytes, and extracellular matrix, an aqueous environment that contains proteins and fibers. In humans, connective tissues constitute about two-thirds of the body. Virchow's observations of connective tissues were crucial to his conceptions of both normal development (cell theory) and cancers' development (connective tissue theory).

In a lecture that Virchow gave at the Institute of Pathology in Berlin in 1858, he explained that the connective tissue is full of germ cells. Virchow defended the idea that all new formations, either normal or pathological, derived from this tissue and its germ cells. This constituted the basis of his connective tissue theory. This theory provides two explanations for the occurrence of cancers. First, as the connective tissue is present throughout the organism, it is capable of producing the same type of cancer in different organs. Second, by their very nature (they are germ cells), connective tissue cells can also produce different forms of neoformations (either in the same organ or in different organs): "We may [consider the] *connective tissue . . . as the common store of germ cells* (Keimstock) *of the body,* and directly trace it as the general source of development of new formations" (Virchow 1860, 398, with some amendments with respect to the Frank Chance translation). To Virchow, as well as Remak and the CSC theory, the cells from which cancers develop have an open differentiation potential and are, themselves, undifferentiated. Thus, the nature of the cells that originate cancers plays a central role in explaining how cancers develop. And, although Virchow came up with his "connective tissue theory" to account for particular cases, like the tibial epithelial cancer he had diagnosed in 1850 or epithelial cancers in meninges, ovaries, and testes, he rapidly applied his theory to all tumors (Virchow 1855b).

Virchow is not the only one to defend this theory of cancers (see Rather 1978, 126). However, anatomists, pathologists such as Carl Thiersch, Wilhelm His, Ludwig Buhl, and Theodor Billroth, who claimed that the laws of embryology apply to any neoformation, pathological or not, criticized the connective tissue theory. Billroth claimed that "a connective tissue cell could no more give birth to an epithelial cell than a dog to a frog." Nevertheless, Edwin Klebs, Theodor Langhans, and Carl Otto Weber, for example, disputed the idea that the laws that govern embryological development also applied to pathologies (reviewed in Rather 1978, chap. 4).

The connective tissue theory is a hypothesis of the stem cell origin of cancers. Similar to both the delocalized embryonic rest theory of Remak and the CSC theory, it attributes the differentiation potential of the cells that originate cancer to their cellular identity, and this, in turn, explains the characteristics of cancer cells. Yet the kind of cells that originate cancers is different for Remak and Virchow: delocalized embryonic rests and germ cells of the connective tissue, respectively.

The Superabundant Rest Theory (Julius Cohnheim)

In the 1870s, Julius Cohnheim adopted the delocalized embryonic rest theory that Remak had proposed. However, he saw no reason to limit the use of that theory to cases of cancers that arose due to embryonic cell delocalization or "detachment."[2] Cohnheim's version of this theory is characterized by two modifications. First, Cohnheim's embryonic rest theory applies to all types of cancers. And second, to Cohnheim, the embryonic rests could either be detached embryonic cells, as suggested by Remak, or be superabundant embryonic cells (i.e., extra cells that the organism did not require to develop). Thus, I propose to call this version of the embryonic rest theory the "superabundant embryonic rest theory."

According to Cohnheim's superabundant embryonic rest theory, embryonic rests in the adult organism are simply extraneous cells that were not used during development. Unlike Remak, Cohnheim did not conceive of embryonic rests as being anatomically delocalized—that is, rests, to Cohnheim, did not belong to a particular organ or tissue type from which they became removed. This change in the conception of the "embryonic rests" enabled Cohnheim to explain every type of tumor development.

> The simplest view appears to me undoubtedly to be that in an early stage of embryonic development more cells are produced than are required for building up the part concerned, so that there remains unappropriated a quantity of cells—it may be very few in number—which, owing to their embryonic character, are endowed with a marked capacity for proliferation. (Cohnheim 1889, 760)

It is worth noting that Cohnheim gives much more weight to what he called the "inherent disposition" of embryonic rests than to the fact that they are superabundant or delocalized. This inherent disposition—the developmental abilities intrinsic to the embryonic cells—is what allows him to explain tumors' development.

> The new-born infant brings with it into the world, not the tumour, but merely the superabundant cell-material, and from the latter, if circumstances be favorable, a tumour may grow later on. I wish once more, however, to warn you against adhering too closely to the expression, "superabundant cell-material;" it would perhaps be more correct to speak of a material having an inherent potentiality for subsequent tumour development. For the development of the tumour depends on this power, which for the rest is simply the quality so very commonly manifested in individual instances of inheritance and development. (Cohnheim 1889, 764–765; see also p. 761)

Conclusion on the Embryonic Cell Rest Theory: Three Hypotheses
of the Stem Cell Origin of Cancers

The three theories developed by Remak, Virchow, and Cohnheim are undoubtedly different from the CSC theory. The term "stem cell" does not appear in any of these works. And, while the concept of "embryonic rests" has been construed as equivalent to our modern understanding of a stem cell, the usage to which these authors put "rests" refers to embryonic stem cells, whereas the CSC theory generally refers to adult stem cells.

Despite these differences, it is still worth considering—as CSCs specialists do—that the theories of Remak, Virchow, and Cohnheim are three different theories of the stem cell origin of cancers. These three theories, as well as the CSC theory, refer to a particular category of cells, with specific functional properties, that give rise to cancers. All of these theories provide the same functional description for those cells: they have a significant differentiation potential. Those cells occupy a similarly privileged position in development: they all sit atop the hierarchy of differentiation. Thus, these are cells that give rise to other cells.

From this historical perspective, the permissiveness of the hypothesis of the stem cell origin of cancers becomes evident with respect to two points:

1. "Stem cell," as a concept, can refer to at least three distinct entities: embryonic cells (Remak and Cohnheim), germ cells (Virchow), and adult stem cells (the CSC theory).
2. The hypothesis can be highly specific or highly generalized in the types of cancers to which it refers. To Remak, it only applies to cancers that display cell types in tumors that contradict their location of development. In contrast, to Virchow, Cohnheim, and the CSC theory, it applies to all cancers.

THE INFLUENCE OF THE EMBRYONIC REST THEORY IN THE TWENTIETH CENTURY: TERATOMAS AND TERATOCARCINOMAS

During the first half of the twentieth century, the embryonic rest theory (either delocalized or superabundant) was often debated. It constituted an alternative to the emerging theories of parasitic origin or chemical origin of cancers, which met with increasing success. However, the embryonic rest theory, in its diverse forms, rapidly disappeared, with the exception of two remarkable cases: bone cancers called "adamantinomas" and germ cell tumors called "teratocarcinomas." Research conducted on teratocarcinomas throughout the twentieth century reveals the historical unity between the embryonic rest theories and the CSC theory. In particular, two researchers, Leroy Stevens and Gordon Barry Pierce, are responsible for the transition between these theories, by showing that the cells that initiate cancers (the "embryonic rests," according to the embryonic rest theory) are "stem cells."

Adamantinomas constitute a fascinating extension of the problem pathologists encountered in the nineteenth century when they handled cancers that conflicted with the germ layer theory. Adamantinomas are cancers made up of epithelial cells that resemble cells belonging to the enamel organ—a structure found in developing teeth.[3] Adamantinomas generally affect the jaw but can also be found in the pituitary gland (pituitary adamantinoma) and in bones, especially the tibia (tibial adamantinoma). The presence of adamantinomas in these three places raised questions about the origin of the epithelial cells that compose them. This is the same problem as the one that pushed Remak to suggest the delocalized embryonic rest hypothesis—namely, how can we explain the occurrence of such cancers without a source of epithelial cells? This question made the delocalized embryonic rest hypothesis relevant throughout the twentieth century, and it is still under investigation.[4]

Despite the persistence of the delocalized embryonic rest theory, the CSC theory has yet to be used to explain adamantinomas. Therefore, the remainder of this chapter will focus on teratocarcinomas, research on which is relevant to both the history of the concept of embryonic stem cell as well as the history of the concept of cancer stem cell.

The Blastomere Theory vs. the Germ Cell Theory

Teratomas and teratocarcinomas are mainly found in the testes and ovaries and are characterized by a great heterogeneity in the cells composing them. Both have differentiated cells from all three germ layers (endoderm, mesoderm, and ectoderm) and, in the case of teratocarcinomas, cells that resemble the undifferentiated pluripotent stem cells found in embryos. The distinction between teratoma and teratocarcinoma is recent, and many authors have used the terms interchangeably.[5] Thus, to avoid confusion, I systematically refer to them as "teratomas/teratocarcinomas."

Max Wilms (1867–1918) made the first histological descriptions of teratomas/teratocarcinomas (called "embryomas" because of their resemblance to embryos) in 1895. To explain the varied histological characteristics of these cancers, Wilms first adopted the germ cell theory developed by Felix Marchand in the 1890s. According to this theory, teratomas/teratocarcinomas come from eggs that escaped oogenesis. However, Felix Marchand rapidly developed, with Robert Bonnet, another hypothesis of the origin of teratomas/teratocarcinomas, called the blastomere theory (Blastomerentheorie)—a hypothesis similar to Cohnheim's superabundant embryonic rest hypothesis and which aimed to explain the histological characteristics of these types of tumors. According to the blastomere theory, teratomas/teratocarcinomas come from "blastomeres." Blastomeres are embryonic cells produced during the cleavages following fertilization—well before the formation of the germ layers—and thus are not restricted in their differentiation capabilities like later stage cells (Figure I.1A). Blastomeres, then, were posited to escape their normal developmental trajectory and could then give rise to teratomas/teratocarcinomas. The great differentiation potential of the blastomeres thus explained the presence of cells in teratomas/teratocarcinomas derived from each of the three germ layers, as well as the presence of clusters of cells that resemble the blastocyst, an early stage of the embryo development (Figure I.1B) (see Maehle 2011).

The germ cell theory and the blastomere theory, which are still the two principal explanations for the origin of teratomas/teratocarcinomas, are functionally equivalent (Martin 1975). In both cases, the developmental potential of the cells that initiate the tumor is invoked to explain tumor cells' characteristics. In other words, these are two alternative and conflicting theories of the stem cell origin of cancers.

At the turn of the twentieth century, both the blastomere and germ cell theories met with increasing interest inside and outside of Germany. Many researchers tried to test embryonic cells' abilities to produce tumors by transplanting embryonic tissue into adult animals. These tests first proved largely inconclusive—most transplanted materials were absorbed by the host organism after some initial growth. The only scientist who managed to grow tumors, and specifically teratomas/teratocarcinomas, through transplantation techniques was Max Askanazy (1865–1940). Askanazy, a professor of pathology at the University of Königsberg (and, later, University of Geneva), described the results of his transplantation experiments in 1907 as "the most 'beautiful' illustration of Cohnheim's theory" (Askanazy 1907, 47; Maehle 2011, 9).

Debates over the Origin of Teratomas/Teratocarcinomas

In the early 1900s, John Beard (1858–1924), an embryologist at the University of Edinburgh, developed the "trophoblastic" theory of cancers—a particular interpretation of the germ cell theory—as an alternative to the blastomere theory. The trophoblast is the outer layer of cells of a blastocyst (Figure I.1B), which develops into the placenta and is involved in the egg's implantation within the wall of the uterus. According to the trophoblastic theory, cancer arises when "irresponsible trophoblasts" derived from "persistent primary germ cells" become "vagrant" (for cancers not localized in ovaries or testes). The germ cells' sudden ectopic development produces neoplasms, similar to the trophoblast (Beard 1900, 478; Beard 1902, 1905a, 1905b).

Beard's trophoblastic theory contributed to animated debates over the origin of teratomas/teratocarcinomas. In 1929, Gilbert William de Poulton Nicholson, for example, stressed that teratomas/teratocarcinomas do not display some characteristics, like membranes, segmentation, and organized vasculature, which are essential to the embryo (de Poulton Nicholson 1929). French physician and biologist Albert Peyron disagreed with de Poulton Nicholson's assessment of the theory. Peyron described a case of a sacrococcygeal embryoma in which he found a "pituitary region with two gland lobes and a vestigial pharynopituitary tract." He claimed that the presence of these structures within the embryoma constituted evidence "for a median embryo axis corresponding to the notochord or to the pre-chordal plate"—structures that would indicate that the embryoma did, indeed, have structural characteristics in common with embryos. Peyron also claimed that he found evidence for "a real lumbar region with its fetal structure" and, above all, for the presence of the "two testicles" within the embryoma. He interpreted these tumors as "real twins," which had undergone irregular and pathological development (yet another version of the hypothesis of the stem cell origin of cancers) (Peyron 1939).

Peyron's argument against de Poulton Nicholson did not end the critique of the hypothesis of the origin of teratocarcinomas in delocalized germ cells, or "vagrant" germ cells, as Beard called them. Emil Witschi (1948), and then Duncan Chiquoine (1954), challenged the *existence* of delocalized primordial germ cells. Witschi, a specialist in sexual differentiation in vertebrates, observed in the collection of human embryos at the Carnegie Embryological Laboratory, in Baltimore, that the germ cells that did not accomplish their migration were eventually reabsorbed. Moreover, his work on sexual differentiation made him skeptical with respect to the ability of germ cells to differentiate into cell types other than ova and spermatozoa (see Maehle 2011, 11–12). Meanwhile, Chiquoine, a researcher at Princeton University, stained primordial germ cells to characterize their migration pathway in the mouse embryo and found no evidence of delocalized germ cells.

> Although a few germ cells were recognized in some embryos as having strayed from the germ-track, the location of these lost germ cells was never too far from the track. Embryomata, on the other hand, are known to occur in almost any part of the body and there are no observations to support such a widespread distribution of lost germ cells. (Chiquoine 1954, 142)

Meanwhile, Elizabeth Fekete and Mary Ann Ferrigno at the Jackson Laboratory in Bar Harbor, Maine, provided evidence against the hypothesis of the germ cell origin of teratomas/teratocarcinomas. Their chromosomal analyses of a serially transplanted teratoma/teratocarcinoma showed that the cells were diploid, which "rules out the possibility of haploid parthenogenetic development of this tumor from an ovum or polar body" (Fekete and Ferrigno 1952, 440). The fact that they were able to serially transplant this tumor without losing the heterogeneity of the cells that compose the tumor also led Fekete and Ferrigno to two conclusions: first, that tumors had pluripotent cells, able to regrow the tumor at each transplantation, and second, that "the varied elements could be produced only by the continuous growth and differentiation of transferred pluripotent cells" (Fekete and Ferrigno 1952, 440).

In 1961, biologist Beatrice Mintz, a researcher at the Institute for Cancer Research (now the Fox Chase Cancer Center), in Philadelphia, showed that in mice and chicks, a "substantial" number of primordial germ cells could get lost and still survive during migration from the yolk sac to the germinal ridge, invalidating Witschi's and Chiquoine's critiques (Mintz 1961). Based on this observation, it again became possible to support the hypothesis of the germinal origin of teratocarcinomas (e.g., Simson, Lampe, and Abell 1968).[6]

This long-lasting controversy over whether germ cells could form teratomas/teratocarcinomas highlights how indeterminate the hypothesis of the stem cell origin of cancers is. Several theories can instantiate the general hypothesis of the stem cell origin of cancers because the concept of stem cell, in this general hypothesis, is referentially open; that is, the different versions of the hypotheses of the stem cell origin of cancers diverge with respect to the stem cell type at the origin of those cancers. The CSC theory does not currently end such indetermination (see Chapter 5).

EMBRYONAL CARCINOMA CELLS, EMBRYONIC STEM CELLS, AND CANCER STEM CELLS

After the 1950s, Cohnheim's superabundant embryonic rest theory, Marchand's and Bonnet's blastomere theory, Max Wilms's primitive germ cell theory, and Beard's trophoblast theory were widespread but not subject to a great deal of investigation. However, research did continue on the developmental origins of teratocarcinomas. This research, specifically the investigations of Leroy Stevens and Gordon Barry Pierce, deserve particular attention because of their contributions to the emergence of the modern conception of embryonic stem cells and the cancer stem cell theory, respectively.

Leroy Stevens's Transplantation Experiments

In the early 1950s, Leroy Stevens, a researcher at the Jackson Laboratory, discovered a mouse line (strain 129) in which males develop testicular teratomas/teratocarcinomas at abnormally high frequencies. Stevens's work with strain 129 helped to revive the debate between the embryonic and the germ cell hypotheses of the origin of cancers (Stevens and Little 1954).

The Jackson Laboratory was established in 1929, with the objective of developing mouse lines to study cancers (see Rader 2004). In 1953, geneticist Clarence Cook Little hired Stevens, who trained as an embryologist, as a postdoctoral researcher with funding from the tobacco industry. Stevens was directed to work on establishing the toxicity of cigarette paper, in order to deflect claims about the toxicity of tobacco (see Lewis 2001, chap. 6). While systematically studying dozens of mice for this project, Stevens noticed a high occurrence of teratomas/teratocarcinomas in mouse strain 129. While teratomas/teratocarcinomas are uncommon in humans, they are extremely rare in mice. Strain 129, in which approximately 1 percent of males develop tumors, provided a valuable model organism on which researchers could conduct systematic research on teratomas/teratocarcinomas. Stevens quickly committed himself to this new research on teratomas/teratocarcinomas, which eventually led him to the question of the developmental origins of these tumors.

In 1941, Elizabeth Jackson and Austin Brues, researchers at the Huntington Memorial Hospital of Harvard University, published the results of a series of teratoma/teratocarcinoma transplantation experiments. The pair used histological characteristics to separate the cell populations of a teratoma from the ovary of a C_3H strain mouse and then transplanted the cells into other C_3H strain mice. Their results showed that (1) only the undifferentiated cells, which look like embryonic cells, were able to produce new teratomas/teratocarcinomas and (2) that cell divisions seemed to occur preferentially, if not exclusively, in those undifferentiated cells (Jackson and Brues 1941). Nine years later, Elizabeth Fekete and Mary Ann Ferrigno (1952) used Stevens's strain 129 mice to transplant teratomas/teratocarcinomas at the Jackson Laboratory and confirmed Jackson's and Brues's observations. This led Fekete and Ferrigno to hypothesize that those undifferentiated cells were pluripotent.

In the course of his work with strain 129 at the Jackson Laboratory, Stevens became interested in the undifferentiated cells that

Jackson and Brues (1941) and Fekete and Ferrigno (1952) had described, as well as in the characterization of the developmental stages of teratomas/teratocarcinomas. When those tumors were transplanted, they gave rise to embryoid bodies similar to embryos in the stage of a blastocyst (Stevens 1959). He observed that those embryoid bodies could then renew themselves and produce differentiated cells from the three embryonic layers (Stevens 1960). Following this characterization of teratoma/teratocarcinoma development, Stevens became interested in their origin, asking the following question: from which cells of the organism do teratomas/teratocarcinomas develop?

Stevens favored the hypothesis that teratomas/teratocarcinomas developed from germ cells. Because he had observed that in mice, teratomas/teratocarcinomas begin to develop in the embryos' testes between 12 and 17 days postfertilization, he decided to transplant the genital ridge of 12-day-old mouse embryos—the embryonic structure in which primordial germ cells develop—into the testes of adult mice to test the germ cell hypothesis. Stevens conducted two series of transplantations: one from sterile mouse embryos (a lineage with a mutation of the *Steel* gene) and the other from mouse embryos with normal fertility. His results showed that transplantations from fertile mice generally led to the development of teratomas/teratocarcinomas (in 75 percent of cases). In contrast, only 2 of the 75 transplantations conducted from sterile mice led to the development of tumors. As sterile mice had little to no germ cells, these results suggested that primordial germ cells (and not supporting cells that constituted the rest of the genital ridges) were responsible for the origin of teratomas/teratocarcinomas (Stevens 1964, 1967).

In the late 1960s, Stevens started graft experiments of entire embryos. He first transplanted fertilized strain 129 single- and two-cell ova (Stevens 1968) and then three-day-fertilized and six-day-fertilized eggs into the testes of adult mice (Stevens 1970). These transplantations sometimes resulted in the proliferation of undifferentiated embryonic cells and sometimes resulted in the occurrence of teratomas/teratocarcinomas, which Stevens was able to maintain and to effectively transplant again. This unexpected result led to two important consequences. First, it led Stevens to reconcile the hypothesis of the origin of teratomas/teratocarcinomas in germ cells and the hypothesis of their origin in embryonic cells. Second, it established a field of research, which eventually led to the production of cultured pluripotent embryonic stem cells (ES cells).

Reconciling Hypotheses of the Origin of Teratomas/Teratocarcinomas in Embryonic Cells and in Germ Cells

In Stevens's work, the occurrence of teratomas/teratocarcinomas from the transplantation of cells belonging to the genital ridge showed that those tumors could originate from primordial germ cells. The occurrence of teratomas/teratocarcinomas from the transplantation of embryos at an early developmental stage (up to the blastocyst) showed that they could originate in embryonic stem cells. Thus, Stevens concluded that those two hypotheses were not contradictory. Both could be true.

The importance of the functional role attributed to the cells from which cancers develop, as well as the close relationship between this role and the cells' intrinsic properties, is very clear in Stevens's work. These cancer-causing cells are "undifferentiated," "pluripotent" cells able to self-renew. They are the only tumorigenic cells—the cells from which all other cells present in teratomas/teratocarcinomas come. Therefore, Stevens supported a model of cancer development that was very similar to the CSC theory. The consequence of Stevens's reconciliation of the hypotheses of the embryonic and germinal origin of cancers is instructive; it shows that the functional properties that define stem cells (self-renewal and differentiation) are not sufficient to precisely identify a referent (i.e., the cell type). Several cell types fall within this functionally defined category—minimally germ cells and embryonic cells.

From the Study of Embryonal Carcinoma Cells to the Isolation of ES Cells

The work of Stevens is not only part of the history of the CSC theory but also of the history of embryonic stem (ES) cells. In his search to determine the cellular origins of tumors in strain 129 mice, Stevens developed a system to culture the undifferentiated cells, called "embryonal carcinoma" (EC) cells, in teratomas/teratocarcinomas. In 1967, Finch and Ephrussi complemented Stevens's technique with *in vitro* culturing protocols (Finch and Ephrussi 1967). By combining these techniques, researchers were able to reveal a great similarity between EC cells and the cells belonging to the inner cell mass of the embryo at the blasto-cyst stage (Martin 1975; Jacob 1977).

Stevens's transplantation experiments had already shown that embryonic cells could act as EC cells because, once transplanted into mouse testes, the embryonic cells could grow teratomas/teratocarci-nomas. The new ability to culture EC cells allowed researchers to inves-tigate the ability of the EC cells to behave like embryonic cells of the inner cell mass. Several teams transplanted cultured EC cells into embryos at the blastocyst stage to see if they would act within a devel-opmental context like embryonic cells (Brinster 1974; Papaioannou et al. 1975). Among those researchers, Beatrice Mintz and Karl Ill-mensee were able to produce chimeric embryos that contained tissues derived from EC cells from Stevens's mice, showing the ability of the EC cells to act like the embryonic cells in the development of the mouse (Mintz and Illmensee 1975).

The next step was then to isolate and culture the embryonic cells of the inner cell mass of the blastocyst embryo. Beginning in the 1960s, several attempts were made to culture the blastocyst's cells and to dem-onstrate their pluripotency—each of these attempts failed (see Sherman 1975a, 1975b). In 1981, Martin Evans and Matthew H. Kaufman at the University of Cambridge and Gail Martin at the University of California in San Francisco finally succeeded in extracting and culturing the cells of the inner cell mass of mouse embryos (Evans and Kaufman 1981; Martin 1981). Once transplanted into mice, they gave rise to teratomas/terato-carcinomas. When cultured without feeder cells, they formed embryoid bodies. Martin called these cultured embryonic cells ES cells "to denote their origin directly from embryos and to distinguish them from embry-onal carcinoma cells derived from teratocarcinomas" (Martin 1981, 7635). It took nearly two more decades before James Thomson and col-leagues, at the University of Wisconsin, Madison, succeeded in cul-turing human ES cells (Thomson et al. 1998).[7]

Gordon Barry Pierce's Contribution

Gordon Barry Pierce also took part in this research on embryonal carci-noma cells. His work with Stevens, in the Jackson laboratory, in 1967, on characterizing the EC cells demonstrated their pluripotency at the single cell level. Additionally, he developed a theory, similar to the CSC theory, which took into account the development of teratomas/teratocarcinomas, as well as cancer development more generally.

Pierce was a physician who joined Frank Dixon's team at the University of Pittsburgh in 1955 to study teratomas/teratocarcinomas. Dixon had already published on testicular cancers with his colleague Robert Moore from the Medical School of the University of Washington. Their research had led them to hypothesize that cancers originate in germ cells. Their argument was as follows:

1. Tumors show "a multipotentiality that approaches that of the germ cell itself."
2. These tumors, which show "both somatic and trophoblastic differentiation should arise from the only cells having such potentialities."
3. "The occurrence of a great preponderance of germinal (or teratoid) tumors in the gonads of either sex further supports the postulated germinal origin" (Dixon and Moore 1953, 429).

Their classification of cancers connected all types of teratomas/teratocarcinomas to an origin in the germ cells (see Dixon and Moore 1953, 437, fig. 13).

Pierce began his career as a biologist, investigating the hypothesis that some cancers (teratomas/teratocarcinomas) originate from cells with a high developmental potential (embryonic cells) (e.g., Pierce and Dixon 1959a, 1959b). In his earliest articles, he explicitly connected his work with that of Julius Cohnheim, from whom Pierce adopted the hypothesis that cancers develop from embryonic cells. Additionally, Pierce saw his own work as a continuation of the research traditions of Max Askanazy, Max Budde, and Rupert Allan Willis. Askanazy was the first person to experimentally confirm that embryonic cells are capable of producing tumors, while Budde was the scientist responsible for the embryonic rest hypothesis of teratomas/teratocarcinomas (Budde 1926). Finally, Willis was the researcher to whom Stevens had attributed the hypothesis that teratomas originate from totipotent embryonic cells (Willis 1958, in particular chap. 11, pp. 442–444). Pierce also drew inspiration from the teratocarcinoma transplantation works of Elizabeth Fekete, Mary Ann Ferrigno, Elizabeth Jackson, and Austin

Brues, whose collected efforts demonstrated that teratomas/teratocarcinomas grow "from foci of embryonal cells which divide rapidly throwing off some cells which retain the ability to grow indefinitely" (Jackson and Brues 1941, 498; Fekete and Ferrigno 1952).

Pierce, like Stevens, belonged to the research tradition in which the hypothesis of the stem cell origin of cancers played a dominant role. Yet Pierce diverged from his contemporaries and predecessors in a number of ways, and his adoption of the stem cell hypothesis of the origin of teratomas/teratocarcinomas is tightly linked to his fight against three dogmas of his time.

First, Pierce was opposed to the idea that carcinogenesis occurs through dedifferentiation and that cancer cells are unable to differentiate. The undifferentiated character of the cancer cells led biologists to the conclusion that the cancer cells have lost their features of differentiation. On the contrary, according to Pierce, "what has been interpreted as dedifferentiation is in reality an abortive attempt at differentiation by the neoplastic stem cells" (Pierce 1970, 1248; see also Pierce 1974, 104). Second, he disagreed with the somatic mutation theory of cancer, according to which cancers are provoked by genetic mutations accumulating in somatic cells. He claimed that cancers are instead produced by epigenetic mutations. He hypothesized that "differentiated cells respond to carcinogens in a manner analogous to embryonic cells responding to embryonic inductors. In other words, carcinogenesis may be explained as a stable response to inductive phenomena acting through cytoplasmic nuclear controls on an initially unaltered genome rather than as an accident of genetic coding" (Pierce 1967, 245). Third, Pierce fought against the proverb, "once a cancer cell, always a cancer cell" (Pierce 1975, 3). Against this dogma, he claimed that differentiation could lead to the production of benign cells. He saw Mintz and Illmensee's chimeric mice produced by transplantation of EC cells in the blastocyst as a major argument in his favor, since those experiments showed that cancer cells could give rise to the non-pathological development of a mouse.[8]

These fights, which characterize Pierce's theory of carcinogenesis, are strikingly dissociated from the context of the current CSC theory. The first dogma—that carcinogenesis occurs through dedifferentiation—

no longer stands. In contrast, the current dogma is that stem cells' differentiation is irreversible. Therefore, the proponents of the CSC theory do not have to fight for the idea that cancer cells can somehow differentiate. The second idea—that cancers occur through somatic mutations—has been incorporated into the CSC theory. Finally, the question of whether or not a cancer cell has the capacity to revert to a non-cancer cell is not addressed by the CSC theory.

Despite these disjunctions with the current CSC theory, Pierce developed a stem cell theory of cancer more similar to the modern CSC theory than any of his predecessors. Pierce was the first person to speak about teratoma/teratocarcinoma stem cells in an equivalent sense to the modern notion of CSCs. His work on teratomas/teratocarcinomas demonstrated cancer cells' differentiation capacities—an uncommon idea in cancer research at the time and one that met with a great deal of skepticism. In 1959, Pierce coauthored an article with Dixon that showed that embryonic carcinoma cells were multipotent and could differentiate—the article remained under revision at the journal *Cancer* for six months, an uncommonly long time. In an interview led by Juan Arechaga, Pierce recalled that one of the associate editors finally telephoned him to say that there was nothing wrong with the data but that they could not publish them because "everybody knows that cancer cells cannot differentiate" (Arechaga 1993, 10). In response to pressure from the journal's editor, Pierce was obliged to change the title of his article from "Teratogenesis by Differentiation of Multipotential Cells" to "Teratogenesis by Metamorphosis of Multipotential Cells" (see Pierce and Dixon 1959a). Through his work to establish the differentiation potential of cancer cells, Pierce ultimately demonstrated that teratocarcinomas' embryoid body cells were multipotent (Pierce, Dixon, and Verney 1960). Pierce's move to the cellular level was an improvement over the work that Stevens had conducted in 1960, in which he had shown that embryoid *bodies* were multipotent (Stevens 1960). Pierce further refined the issue of differentiation potential when, in 1964, he provided evidence for the fact that cells composing embryoid bodies were pluripotent. Along with his student, Lewis Kleinsmith, they were able to dissociate embryonic carcinoma cells and to obtain clonal lineages from a single cell. This permitted them to demonstrate that multipotency was not just a population feature, but it also applied at the single cell level (Kleinsmith and Pierce 1964). His work on cancer cells led Pierce to adopt a hierarchical model of teratomas/teratocarcinomas, according to which these cancers would develop from pluripotent stem cells, which in turn would give birth to the "differentiated structures of the tumor" (Pierce and Verney 1961).

Pierce rapidly expanded his hierarchical model toward a more general model of cancer development (Pierce 1967, 1974, 1977b; Pierce et al. 1977; Pierce, Shikes, and Fink 1978; Sell and Pierce 1994). It was within this context of grappling with cancer theory that he introduced the concept of "neoplastic stem cell" (Pierce 1977a) whose referent is close, if not identical, to the concept of cancer stem cell in the CSC theory. The neoplastic stem cell is a stem cell capable of self-renewal and differentiation, just like CSCs, and belongs to the tissue in which cancer develops. Thus, they are embryonic or germinal cells in the case of teratomas/teratocarcinomas and adult stem cells in the case of non-embryonic cancers. Additionally, just like for the CSC theory, Pierce's theory relied on an analogy with the development of healthy/normal tissue: "Malignant tissue, like normal tissue, maintains itself by proliferation and differentiation of its stem cells" (Pierce and Johnson 1971, 140; see also Pierce 1974). Finally, Pierce claimed that "the target in carcinogenesis is the normal stem cell of the particular tissue" (Pierce 1970, 1254). Thus, despite the contextual differences between the theory proposed by Pierce and the modern CSC theory, the two are, conceptually, very close.

While Pierce's work advanced the field's understanding of both stem cells and the origin of cancers, there are some issues in his conception and application of stem cells that bear review. First, Pierce repeatedly uses the word "stem" to refer to the original cell of a cancer, regardless of its stemness—Pierce's application of the term "stem" is thus ambiguous with regard to current usage. Second, Pierce's theory of cancer development does not seem to be contingent on stem cells as the source of cancer—thus, to Pierce, cells that initiate cancers need not be stem cells.[9] Indeed, during the interview he gave to Juan Arechaga in August 1991, Pierce claimed that, even if those cells were not stem cells, this would not affect his conception (Arechaga 1993). This position puts Pierce at odds with the CSC theory.

CONCLUSION

The idea that cancers would originate in stem cells or in embryonic cells is as old as cell theory. The delocalized and superabundant embryonic rest theories (Remak and Cohnheim, respectively), the connective tissue theory (Virchow), the primitive germ cell theory (Wilms, Beard, and Stevens), and the blastomere theory (Marchand, Bonnet, and Askanazy) are all very different from the CSC theory. However, they are all hypotheses of the stem cell origin of cancers in the sense that all of them assume that cancers develop from a particular population of cells, which are characterized by a high developmental potential. Therefore, this historical overview shows the great permissiveness of the hypothesis of the stem cell origin of cancers.

The diversity of theories of stem cell origin of cancers up to the CSC theory pushes us to wonder what, within this theory, the concept of stem cell refers to. It also pushes us to question its logical extension: does the CSC theory apply to all cancers? If it applies to all cancers, does it do so in the same way? And, is the stem cells' referent different between cancers?

4

LEUKEMIC AND HEMATOPOIETIC STEM CELLS

The previous chapter highlighted the historical roots of the CSC theory for solid cancers since the mid-nineteenth century. The CSC theory also has several historical roots for blood cancers (leukemia). The notion of "leukemic stem cell" emerged around the 1970s, following from three different lines of investigation:

1. Research surrounding the clonal origin (from a single cell) or multicellular origin (from many cells) of cancers
2. Research on cell division within cancer
3. The development of bone marrow transplantation procedures

These historical roots highlight the ambiguity of the notions of CSC and stem cell regarding their referents: which cells are stem cells? Which leukemic cells are leukemic stem cells?

In the mid-twentieth century, cancer research surrounded a debate about the origin of cancers—namely, do cancers originate from the transformation of one or many cells? Biologists framed two hypotheses: the clonal origin of cancers and the multicellular origin of cancers. According to the hypothesis of the "clonal" origin, cancers would develop from a single cell. The progeny of this original cell would constitute a "clone" or a "clonal population." In contrast, the "multicellular" hypothesis understood cancers to be derived from the transformation of several cells.

In the 1970s, this debate led to the codiscovery that different lineages of cells, in both leukemia and normal hematopoiesis, come from a single kind of cell, which were called stem cells.

*The Clonal Origin Hypothesis and an Explanation
of Cancer Cell Heterogeneity*

In the late 1960s/early 1970s, biologists investigating the origins of cancers, particularly leukemias, proposed the hypothesis that cancers originate from a particular kind of cancer cells—namely, stem cells. During the first half of the twentieth century, the issue of clonal or multicellular origin of cancers had emerged from cytological studies. In the 1950s, research on chromosomes, made possible due to photomicrography and *camera lucida* drawings, led to the observation that cancer cells' chromosomes vary in number. This observation immediately led to questions about the origin of cancers, such as, is chromosomal heterogeneity between cells the result of an evolutionary process (i.e., cellular mutations and subsequent inheritance)? Or were these chromosomal differences present when the cancer originated (e.g., Sato 1952; Makino and Kano 1953; Makino 1956, 1959)?

Biologists have used several methods to answer whether cancers develop from one or many cells, including observing the following:

- chromosomal abnormalities associated with leukemia, like Philadelphia chromosome (which results from a translocation between chromosomes 9 and 22), the loss of a chromosome 7, or the presence of a third chromosome 8 (Tough et al. 1963; Whang et al. 1963; Rastrick 1969; Keinanen et al. 1988);

- natural markers, like the enzyme glucose-6-phosphate-dehydrogenase (G-6-PD) that can have different enzymatic forms (Linder and Gartler 1965, 1967; Gartler et al. 1966; Murray, Hobbs, and Payne 1971; Fialkow et al. 1978; Jacobson et al. 1978), or the immunoglobulins, small proteins that have different forms due to rearrangements of the heavy chain (Fialkow et al. 1973; Thiel et al. 1976, 1977; Wetter, Delbruck, and Linder 1978; Kubagawa et al. 1979);
- and artificial markers such as recombinant DNA, a DNA molecule created in the laboratory and injected into cells (Fearon et al. 1986).

These methods all consist of identifying a cellular or molecular marker, which permits the scientist to follow cell lineages. By so doing, they can investigate the clonal or multicellular origin of cancer, using the following logic:

> A neoplasm that has a clonal origin begins, by definition, in one cell (e.g. in an A cell), and thus all cells in that tumor will have one type (A) as descendants of the one A progenitor cell. If, in contrast, a tumor is found to contain neoplastic cells of both A and B types, it must have had a multicellular origin. (Fialkow 1979, 135)

Results from the observation of those markers all converged toward the idea that leukemia originates from a single cell (clonal origin hypothesis). Whatever the marker that was chosen, it was homogeneously present within the cancer cells. This demonstration raised a new question: how can we explain the heterogeneity of the cells in cancers via clonal origin?

Biologists suggested three hypotheses to account for cellular heterogeneity under the assumption of a clonal origin of cancer. First, that "a developing malignancy could incorporate elements from other lesions, and hence become 'heterogeneous'" (Heppner and Miller 1983, 9). Experimental results showing that the G-6-PD enzyme form is homogeneous among cancer cells make the validity of this hypothesis unlikely. G-6-PD is an enzyme that can have different forms. It is produced from a gene on the X chromosome. Because of the early random inactivation of one X chromosome in each cell of the embryo (known as the Lyon principle), XX heterozygous individuals display a natural mosaic. For those individuals, one would expect that recruitment of other cells by the cancer would also bring in cells with alternative forms of G-6-PD. Homogeneity of G-6-PD in cancer cells of heterozygous individuals makes the incorporation hypothesis highly unlikely. The second hypothesis, introduced by Peter Nowell (1976), involved genetic instability: cancer cells are subject to accumulation of mutations, which leads to the production of heterogeneous subclones with different mutations (see Chapter 2). The third hypothesis, also known as the stem cell hypothesis, was that the common cell from which the cancer cells originate are undifferentiated cells with intrinsic abilities to produce multiple cell types (leukemic stem cells).

Studies on the clonal origin of leukemia greatly affected research on the hematopoietic system. Specialists in the hematopoietic system dealt with a question that was similar to the clonal or multicellular origin of cancers—namely, do all of the different blood cells come from the same progenitor, i.e., a hematopoietic stem cell (e.g., Lichtman 2001)? For example, at the end of the 1950s, David Barnes and colleagues, at the Medical Research Council Radiobiological Research Unit, in Harwell, England, studied the recovery of irradiated mice following bone marrow and spleen transplantation and observed that the same chromosomal rearrangements, caused by irradiation, were present in myeloid and lymphoid cells, which led them to argue in favor of the "monophyletic theory of hematopoiesis," according to which "all cells of the myeloid and lymphoid series stem from a single progenitor" that they refer to as the "hemocytoblast" (Barnes et al. 1959, 7). The results obtained with leukemia also provided strong arguments for the common origin of blood cells, as they showed that several hematopoietic cell types (e.g., granulocytes, erythrocytes, and megakaryocytes) derive from the same cell, a stem cell (see in particular Tough et al. 1963; Whang et al. 1963; Fialkow, Gartler, and Yoshida 1967): "In addition to indicating a clonal origin of this malignancy, our results provide strong support for the hypothesis that erythrocytes and granulocytes have a common stem cell" (Fialkow, Gartler, and Yoshida 1967, 1468). In addition to monitoring the ability of leukemic cells to differentiate into granulocytes (Whang et al. 1963; Fearon et al. 1986), erythrocytes (Blackstock and Garson 1974; Clein and Flemans 1966; Trujillo and Ohno 1963), and megakaryocytes (Tough et al. 1963), biologists also investigated the ability of leukemic cells to differentiate into B and T cells of the immune system (Fialkow, Jacobson, and Papayannopoulou 1977). As a result, in 1980, Philip Fialkow (1937–1996), at the University of Washington in Seattle, distinguished several stem cell types, such as "myeloid stem cells" and "leukemic stem cells," the latter being also described as "pluripotent."[1]

> Most of the chronic hematopoietic proliferations studied with G6PD have been found to arise in pluripotent stem cells.... An exception is CLL [chronic lymphocytic leukemia] which is expressed in cells with differentiation restricted to the B-lymphocyte pathway. The G6PD data in CML [chronic myeloid leukemia] provide evidence for the existence in man both of a stem cell multipotent for the lymphoid and myeloid systems, and for the existence of restricted stem cells committed to differentiate only into myeloid cells or only into T lymphocytes. (Fialkow 1980, 31)

These studies show an early occurrence of the modern idea of cancer stem cell with respect to a particular type of cancer (leukemia). As in the case of cancer stem cell theory, Fialkow argued that leukemias were hierarchically organized and developed from particular cells: cancerous pluripotent stem cells. However, Fialkow's above quotation departs from the current understanding in that he uses the concept of stem cell to refer to diverse groups of cells among which some are not considered stem cells anymore. He and his contemporaries used the concept of stem cell to refer to (a) the multipotent cells that are able to produce cells of the lymphoid and myeloid lineages of the hematopoietic system and (b) cells already committed to a differentiation pathway and that can only produce myeloid cells, for example, (c) cells with even less differentiation potential, such as cells only able to give rise to the T lymphocytes.

What Does the Concept of Stem Cell Refer To?

The concept of stem cell finds its origin in late nineteenth-century Germany. A striking character of the few historical accounts that focus on the early notion of stem cell is the variability of the referent to which the term "stem cell" was applied.

Ernst Haeckel first framed the concept of stem cell *(Stammzelle)* in 1877 to describe the fertilized egg cell as the cell from which arise all other cells of the developing organism. The use of the term "stem cell" rapidly evolved, to refer to many other cells, both in the embryo and in the adult organism. Valentin Haecker (1864–1927) referred to the "stem cell" as the common precursor cell of the primordial germ cells and of the primordial somatic (mesoderm) cells. Theodor Boveri (1862–1915) conceived of stem cells as those cells that derived from the fertilized egg cell and led to the primordial germ cell and from which the various primordial somatic cells branched off. Richard Weissenberg (1882–1974) used the term "stem cell" to refer to early precursor cells of egg cells and sperm cells.

The term was also introduced in the hematopoietic system in the beginning of the twentieth century to refer to the common precursor cell of myelocytes and lymphocytes (Arthur Pappenheim), of erythrocytes and granulated leukocytes (Wera Dantschakoff), and of all the blood cells (Alexander Maximow). Some, like Paul Ehrlich, argued that there were two distinct populations of blood cells, lymphocytes and leukocytes (granulocytes), that come from two distinct populations of stem cells localized in two different places, the lymph nodes (and spleen) and the bone marrow, respectively (Ramalho-Santos and Willenbring 2007; Maehle 2011; Morange 2014). In the 1970s, as highlighted earlier, the term "stem cell" was still used to refer to various precursor stages of the blood lineage, creating confusion about the exact referent of the term. To overcome this state of confusion, biologists debated several distinctions, namely:

1. between "dependent stem cells" and "independent stem cells," which Ernest McCulloch proposed and Laszlo Lajtha criticized (McCulloch, Till, and Siminovitch 1965; Lajtha 1966);
2. between "pluripotent stem cells" and "committed stem cells," particularly apparent in French journals, including *Nouvelle Revue Française d'Hématologie Blood Cells* and *Bulletin du cancer* (e.g., Cronkite and Feinendegen 1976; van Bekkum and Knaan 1978);
3. between "multipotent stem cells" and "restricted stem cells," which Philip Fialkow stressed (Fialkow 1980); and

4. between "stem cells" and "progenitors," which is the distinction the scientific community finally accepted (for a review, see Ebbe 1968).

These distinctions aim to clarify two things about stem cells: what property defines them and what cells they refer to. Only the distinction between "stem cell" and "progenitor" attempts to narrow down the domain over which the concept of stem cell extends: stem cells are only those cells that can give rise to all the blood cell types, and the cells already committed to differentiation toward one lineage are progenitors. The other examples distinguish between subpopulations of stem cells (e.g., pluripotent stem cells vs. committed stem cells). These historical insights about stem cells highlight the following:

1. The genealogical position was crucial to the concept of stem cell: "stem cell" refers above all to a cell that is at the origin of other cells.
2. The concept of stem cell was prone to accommodate cell heterogeneity: different cells can be stem cells.

The same referent pluralism held for cancer research in the 1970s. Some biologists, like Philip Fialkow, used the expression "leukemic stem cell" to refer to the common cellular origin of *all* the hematopoietic lineages, while other biologists used it to refer to the cell that was common to *some* lineages—the ones they observed in particular types of leukemia.

CELL DIVISION IN CANCERS: SELF-MAINTAINING SYSTEM OR CSC COMPARTMENT?

Alongside research on the clonal origin of cancers that paved the way for the concept of "leukemic stem cell" to emerge, observations of cell division in cancers resulted in a debate that involved two main hypotheses of cancers' development that also involved the notion of stem cell in cancer.

Beginning in the 1950s, research conducted by Giovanni Astaldi and Carlo Mauri on hematopoietic (Astaldi, Allegri, and Mauri 1947; Astaldi and Mauri 1950) and leukemic (Astaldi and Mauri 1953) cell division suggested that cancer cells do not divide more than hematopoietic cells.[2] Their surprising findings initiated research on normal and cancerous hematopoietic cell division. The establishment of protocols for the incorporation of radioactive T nucleotides (called tritiated thymidine, H_3T) into DNA strands *in vivo* allowed scientists to mark cells that divide.

Using H_3T incorporation, several groups confirmed the observation that leukemic blast cells divide less than immature hematopoietic cells. Eugene Cronkite and colleagues at the Brookhaven National Laboratory, New York, as well as Felice Gavosto and colleagues, at the General Medical Clinic at the University of Turin, Italy, observed, in the late 1950s and early 1960s, that H_3T incorporation was low compared to normal hematopoietic cells, suggesting a reduced proliferative capacity in acute leukemia cells (Cronkite et al. 1959; Gavosto, Maraini, and Pileri 1960; Killmann et al. 1962). Gavosto reported labeling indices around 50 percent for normal bone marrow blast cells and around 10 percent for the blasts of acute leukemia (reviewed in Gavosto et al. 1967b).

In 1962, Alvin Mauer and Virginia Fisher, at the Children's Hospital Research Foundation and Department of Pediatrics, Cincinnati, Ohio, reported that more leukemic cells of the bone marrow incorporate thymidine than leukemic cells of the blood. To explain this difference, they suggested that either "some leukemic cells lose the ability to divide after several generations, resulting in a population of non-proliferating blast cells" that then leave the bone marrow and circulate in the blood, or "the proliferative inactivity of cells found in the blood may be only a temporary feature, perhaps determined by environmental factors, which may change on return to a more suitable site" (Mauer and Fisher 1962, 1085; see also Killmann 1965).

Sven-Aage Killmann, at the University of Copenhagen, Denmark, made the same observation in 1965. He also formulated two alternative explanations similar to the hypotheses of Mauer and Fisher, except that Killmann framed them around the concept of stem cell. Killman considered stem cells to be "self-perpetuating" cells that "maintain one or more cell lines by supplying cells for differentiation" (Killmann 1965, 276). Based on the idea that "in acute leukemia, the blast cell line must necessarily be maintained by a stem cell pool," Killmann suggested two explanations. In the first explanation, leukemic stem cells directly feed the non-proliferative maturation pool. In the second explanation, the proliferative leukemic cells are not stem cells but a proliferative or multiplicative cell pool that is inserted between the leukemic stem cells and the maturation pool and that "depend on the influx from an unrecognized stem cell pool" (Killmann 1965, 276). Killmann was influenced by the feeder-sleeper stem cell model, according to which two populations of hematopoietic stem cells are distinguished: some are in a dormant state (they are called sleeper cells), while others are actively proliferating (they are called feeder cells). Feeder cells maintain themselves and feed the differentiating cell lines. Sleeper cells produce new feeder clones when they are exhausted. Killmann (1968) used this model to suggest that leukemia originates in sleeper cells.

In 1964, Gavosto et al. observed that the blast cells were heterogeneous in size and that there was a correlation between size and proliferation: division occurs in larger cells (Gavosto et al. 1964). Mauer and Fisher (1966) confirmed this observation. Three years later, Gavosto et al. (1967a, 1967b) determined that the population of non-dividing smaller cells comes from the division of the proliferative larger cell population, and the observation that "after division more than 50% of the daughter cells become small" led them to conclude that "the large blast does not display self-maintaining kinetics" (Gavosto et al. 1967b, 305). Mauer and Fisher suggested two hypotheses to explain the maintenance of the large blast cell population: either "the dividing cells could have arisen from continuing leukemic transformation of normal cells" or "the small, non-dividing cells of blood and bone marrow could once again have enlarged to go through another cycle of cell division. In this latter case the leukemic cells could have been a closed, self-replicating population" (Mauer and Fisher 1966, 443). Gavosto et al. clearly opposed the latter hypothesis, claiming that nonidentified leukemic stem cells must feed the blast cell population, in accordance with Killmann's second stem cell model of an unrecognized stem cell pool:

> Recently, some Authors posed the problem whether or not
> acute leukemia blast cells are stem cells. The kinetic study
> of acute leukemia blast cells and, specifically, of the fraction
> of blasts which have a proliferative activity similar to that of
> normal blasts, gives clear evidence that blast cells do not

behave as stem cells from a kinetic point of view, namely, that the proliferating blast compartment is not self-maintaining but needs to be fed from another precursor compartment. (Gavosto et al. 1967b, 305–306; see also Gavosto et al. 1967a)

Bayard Clarkson and colleagues at the Sloan-Kettering Institute for Cancer Research in New York tested longer infusions of H_3T in patients with acute leukemia (8–10 days and 20–21 days). In the late 1960s, these longer infusions led them to two observations: (1) at 20 days, almost all cells were labeled (92–99 percent), and (2) small cells can enlarge and divide. They concluded that interconversions between cells of different sizes are possible and that the dormant cells have "the potential to resume proliferation" (Clarkson et al. 1970, 1255). There-fore, in opposition to Gavosto et al., they argued in favor of the hypoth-esis that acute leukemia is a self-maintaining system. In 1971, Clarkson and his colleague Jerrold Fried published a more precise version of this hypothesis wherein they calculated that according to the available data, "during relatively early disease, a total of 10 billion leukemic cells are as-sumed to be present but only about one fifth of these are stem cells, and of the latter, only 65 per cent are assumed to have unlimited proliferative potential [uncommitted stem cells, by opposition to committed stem cells that can undergo a maximum of four maturation divisions before dying, according to their calculation]" (Clarkson and Fried 1971, 581).

In these lines of investigation on the division of acute leukemic cells, at least two opposing hypotheses involve the idea that leukemic cells originate from leukemic stem cells. Interestingly, the disagreement between the proponents of each hypothesis is about which leukemic cells are the leukemic stem cells, illustrating once again the ambiva-lence of the referent of the stem cell concept until the 1970s.

THE DIFFERENCES IN CLONOGENICITY AMONG CANCER CELLS

The third line of investigation that led to the notion of leukemic stem cells was research related to bone marrow transplantation that began in the 1950s. Bone marrow transplantations provided hematologists and oncologists with a tool to study the hematopoietic system *in vivo*, leading to the discovery, in the early 1960s, of multipotent hematopoietic cells capable of generating the multiple blood cell types. In the late 1960s and 1970s, development of *in vitro* culture cell techniques and *in vivo* transplantations showed that only a small fraction of hematopoietic cells and leukemic cells has the ability to grow colonies of heterogeneous cell types. Finally, starting in the 1980s, new experimental tools such as immunodeficient mouse models and cell-sorting machines (fluorescence-activated cell sorting [FACS]) led to an increasingly refined characterization of the cells constitutive of the hematopoietic system and to a race to identify the hematopoietic and leukemic stem cells—a race we are still running.

Bone Marrow Transplantation

Several incidents during the mid-twentieth century garnered interest and concern about the effects of radiation. Events such as the Hiroshima and Nagasaki bombings, discovering radioactive fallout in Nevada (United States) and in Semipalatinsk (now Semey, Kazakhstan), the contamination of the *Daigo Fukuryū Maru* in the Bikini atoll during the A-bomb tests, and reactor accidents such as those in Los Alamos (New Mexico), Oak Ridge (Tennessee), and Vinča (Serbia) generated intensive research into treatments and/or protective agents to fight against the effects of radiation. It was in the context of this radiation-oriented research that bone marrow transplantation was developed (Kraft 2009). In 1951, Emma Shelton's team at the National Cancer Institute in Bethesda, Maryland, showed that by transplanting bone marrow from healthy mice into those that had received lethal doses of radiation, the hematopoietic system could recover (Lorenz et al. 1951). Following from Shelton's work, in 1956, Charles Ford and his colleagues at the Atomic Energy Research Establishment in Harwell, United Kingdom, traced cell lineages following bone marrow transplantation and were able to demonstrate that the reconstitution of the hematopoietic system in mouse transplant recipients was due to the proliferation of the donor's bone marrow cells (Ford et al. 1956). Thus, bone marrow transplantation provided hematologists and oncologists with the first experimental protocol to study the hematopoietic system and bone marrow *in vivo*. Furthermore, the development of bone marrow transplantation as a therapeutic tool contributed to the diffusion of the concept of stem cell.

In 1961, James Till and Ernest McCulloch, at the Ontario Cancer Institute in Toronto, framed an experimental system to study *in vivo* transplanted bone marrow cells' proliferative abilities, as well as their capacity to differentiate. Till and McCulloch found that the injection of healthy bone marrow cells into irradiated mice led to the appearance of nodules in the spleen. Those nodules indicated that the injected cells had a high potential for proliferation and differentiation. Through histological examinations of the nodules, Till and McCulloch (1961) were able to isolate several types of blood cells, including erythrocytes, granulocytes, and megakaryocytes.

Two years later, with their graduate student Andy Becker, Till and McCulloch uncovered evidence for the clonal origin of the splenic nodules: by inducing chromosomal anomalies through irradiation, they were able to trace a common origin of the cells belonging to the diverse lineages within the nodule populations (Becker, McCulloch, and Till 1963). Their discovery of the clonal origins of the nodules led them to hypothesize that the multiple cell lines that they found within the nodules must be derived from hematopoietic stem cells that they characterized by three properties: extensive proliferation abilities, differentiation, and self-renewal (Till et al. 1964, 30). Till and McCulloch named the cells that gave rise to the nodules in the spleen CFU-S (colony-forming unit–Spleen). The combination of *in vivo* experimentation and an *in vitro* culturing system quickly became the standard model for research on hematopoietic stem cells, and in 1980, the pair described a hierarchy of hematopoietic cells with specific clonogenic abilities.[3]

In parallel, biologists developed similar protocols to investigate leukemia (Griffin and Löwenberg 1986). By transplanting leukemic cells, they discovered that leukemic cells behaved similarly to hematopoietic cells, i.e., only a small fraction of leukemic cells is capable of producing large populations composed of heterogeneous cancer cells. Researchers specializing in acute myeloid leukemias called these cells "AML-CFU" (acute myeloid leukemia–colony-forming unit). As early as 1969, Donald Metcalf, Malcolm Moore, and Noel Warner, at the Walter and Eliza Hall at the Institute of Medical Research, Melbourne, Australia, hypothesized a comparable organization for leukemia and the hematopoietic system, with a hierarchy of cells deriving from a pool of stem cells:

> In every culture experiment, clearly only a small proportion of tumor cells could form colonies *in vitro*. In part, this was due to the presence of cells in the tumor mass which no longer had the capacity for division, *e.g.*, tumor metamyelocytes and polymorphs. Beyond this, however, it seems clear that some cells in the tumor have a stem cell capacity, expressed here as the capacity to generate colonies of both granulocytic and mononuclear cells *in vitro*. (Metcalf, Moore, and Warner 1969, 995)

While the protocols for *in vivo* experimentation and *in vitro* cell culture and assay developed by Till, McCulloch, and others led to important discoveries about the presence and role of stem cells in the hematopoietic system and leukemia, it had severe limitations with respect to the identification of stem cells. The protocols developed for these research systems were not sensitive enough to allow scientists to perfectly sort stem cells from other multipotent cells within CFU-S and AML-CFU populations (Buick, Minden, and McCulloch 1979; McCulloch and Till 1981; Griffin and Löwenberg 1986, in particular p. 1189; Dick 2008, in particular p. 4797). The heterogeneous clonogenicity of these populations, then, led to doubt about the hypothesis that CFU-S and AML-CFU corresponded to stem cells.

While imperfect in their detection abilities, the protocols that Till and McCulloch developed initiated a race to characterize leukemic and hematopoietic stem cells and provoked scientists to elaborate a hierarchical model of stem cell differentiation (see Fagan 2007, 2010). According to this model, the hematopoietic system and leukemias are organized into hierarchies, with stem cells at the apex of the hierarchy. At each stage, differentiation occurs, causing cells to specialize and lose part of their potential to differentiate into other cell types. Thus, stem cells, as the precursor of all subsequent lineages, would have the highest differentiation potential. The problem with *in vivo* and *in vitro* clonogenic assays remains that they cannot isolate pure populations of stem cells; they isolate heterogeneous populations of cells, which contain stem cells but also a number of progenitors already committed to differentiation.

Gene Therapies, Murine Models, and Monoclonal Antibodies

In the 1990s, isolation of hematopoietic stem cells became a more pressing issue due to the emergence of a new research field in medicine: gene therapy.[4] The possibility of transferring specific genes into cells using retroviral vectors, called transduction, raised the hope of curing certain diseases, including blood disorders such as β-thalassemia and sickle-cell disease, as well as leukemia. The long-term goal of gene therapies in the 1990s was the "permanent correction of genetic deficiencies of the hematopoietic system" (Larochelle et al. 1995, 1329). The "intense competition to be the first group to show hematopoietic stem cell transduction" played a major role in research on hematopoietic stem cells (Dick 2008, 4807). Indeed, to meet the goal of a "permanent correction" of gene-based disorders, it is required "that genes are introduced into pluripotent stem cells, because only these cells can initiate long-term reconstitution of the entire hematopoietic system" (Larochelle et al. 1996, 1329). The development of gene therapies relied on the discovery and isolation of hematopoietic stem cells, something that the *in vivo* and *in vitro* clonogenic assays available in the 1980s had failed to achieve. In addition, the *in vivo* transplantation protocol for mice was not appropriate for clinical research because mice presented strong immune response to grafts from other species (xenografts), despite irradiation. Research in gene therapies has therefore been a driving force in the production of immunodeficient mouse models.

Mouse models to study human cancers were already in use in the 1980s. The line of mutant mice called *nude*, which have no thymus and therefore no T lymphocytes, allowed scientists to conduct xenografts. While the *nude* model was effective for solid cancer transplantations, attempts to get liquid-based cancers, such as leukemia, to propagate within the host's hematopoietic system failed. In the late 1980s/early 1990s, cancer biologists, such as John Dick's team in Toronto, and immunologists specializing in human immunodeficiency virus, such as Darcy Wilson's team in La Jolla, California, quickly produced new lines of more immunodeficient mice, particularly *bg/nu/xid* and SCID mice, which permitted the first transplantations of human leukemia into mice (Kamel-Reid et al. 1989; Dick 1996; Mosier et al. 1988; see also McCune et al. 1988; Fraser et al. 1995).

Transplantations in SCID mice are possible, but they face experimental constraints. In particular, they require an enormous number of cells (greater than 10^7 cells) to be transplanted to escape the residual immune systems of the host mice. Due to this imperfection, many mice's breedings, with mice displaying different mutations affecting their immune systems, have been developed following the aim to produce a better mouse model. In the mid-1990s, the mouse strain NOD/SCID (nonobese diabetic/severe combined immunodeficiency) became prevalent because it was the only one that permitted serial transplantations (Dick 1996). Serial transplantations, which consist of repeatedly transplanting leukemia from one mouse to another, are necessary to demonstrate the long-term self-renewal ability of stem cells. The serial transplantation process works as follows: biologists graft cells from a human leukemia sample into an irradiated NOD/SCID mouse. Once the transplantation takes hold in the host mouse's bone marrow and reproduces the original leukemia, the graft is transplanted into another NOD/SCID mouse. If the grafted cells are still able to develop leukemia similar to the original donor, then one can conclude that in the original sample, there were multipotent stem cells capable of long-term self-renewal. Without stem cells, the transplanted cells' self-renewal ability would run out during the first transplantation, and no secondary transplantation would be possible. The NOD/SCID mouse line permitted the first phenotypic characterization of cancer stem cells in the mid-1990s.

Phenotypic characterization of stem cells became possible in the 1980s thanks to the conjunction of two technologies developed in the 1970s—a cell-sorting technology called FACS and fluorescent monoclonal antibodies (see Cambrosio and Keating 1992b; Keating and Cambrosio 1994). The first prototype of FACS became available in 1974 at Stanford, where Irving Weissman was among the early users, and became commercially available in the 1980s. The FACS technology generates droplets containing single cells that the machine can count and sort based on some properties such as size, granularity, and fluorescence. With the production of monoclonal antibodies tagged with fluorochromes, FACS was able to sort cells according to specific cell surface molecules, called antigens. Monoclonal antibodies are proteins produced by immune cells (B lymphocytes) that recognize and bind to specific antigens. To produce large quantities of monoclonal antibodies, George Köhler and César Milstein (1975), at the MRC Laboratory of Molecular Biology, Cambridge, United Kingdom, came up with the idea to fuse B cells with myeloma cells—cancer cells that are easy to culture. This fusion, for which they received the Nobel Prize in medicine in 1984, resulted in a population of cells called hybridomas. Hybridomas are able to indefinitely reproduce and can thus produce a heightened number of antibodies (see Cambrosio and Keating 1992a, 1995). The utilization of monoclonal antibodies, in combination with FACS, made it possible to sort cells based on their antigens, without killing them.

Weissman's group at Stanford University's School of Medicine was among the first to use the FACS technology, in the late 1970s, to characterize the cells of the hematopoietic system, from the most differentiated back to the hematopoietic stem cells. Other groups across the world began to search for the hematopoietic stem cells in the 1980s, when FACS was commercialized. With FACS and the monoclonal antibodies, hematologists and immunologists were able to sort cells and transplant them into mice to analyze their functional properties. This work enabled researchers to study the differentiation potential and self-renewal capabilities of the blood cells according to their cell surface markers (see Fagan 2007, 2010). In 1988, Weissman's group claimed to have identified the hematopoietic stem cells, which they characterized as Thy-1^{lo}Lin$^-$Sca-1$^+$, in mice. However, this claim was rapidly challenged. The Thy-1^{lo}Lin$^-$Sca-1$^+$ cell population was heterogeneous, and more refinement was needed to characterize pure populations of hematopoietic stem cells (reviewed in Fagan 2007). By 1994, Weissman's team subdivided the population into three subpopulations they called "long-term hematopoietic stem cells," "short-term hematopoietic stem cells," and "multipotent progenitors" (Morrison and Weissman 1994).

At the same time, John Dick's laboratory, based in Toronto, also characterized populations rich in hematopoietic stem cells, using two cell markers: the presence of the antigen CD34 (CD34$^+$) and the absence of the antigen CD38 (CD38$^-$). In 1994, they began to sort cells from human acute myeloid leukemia samples with these markers. After transplantation of the sorted leukemic cells into SCID mice, Dick's group found that only the CD34$^+$CD38$^-$ leukemic cells were capable of producing leukemia. They called these cells "leukemic initiating cells" (Lapidot et al. 1994; Bonnet and Dick 1997). This experiment is widely considered the first evidence for the existence of CSCs and lent huge support for the hierarchical model of cancers' organization and development. The experimental setting was eventually adjusted to solid cancer research and, in 2003, was used to isolate cancer stem cells in the breast (Al-Hajj et al. 2003; Dontu et al. 2003) and brain cancers (Singh et al. 2003, 2004; Hemmati et al. 2003). The concept of cancer stem cell eventually spread within the oncological community, and CSCs were finally isolated in numerous cancers during the 2000s.[5]

A CONCEPTUAL PROBLEM FROM EXPERIMENTAL PRACTICES

The techniques (FACS, monoclonal antibodies, transplantations) that led researchers to frame CSC as a concept in the early 2000s raise an issue about CSCs as well as about stem cells more generally. The concept of stem cell (either cancer or non-cancer stem cell) refers to individual cells. However, experiments to isolate and study those cells are done at the population level. Transplantations require the injection of a large number of cells to overcome the radiation to which immunodeficient mice are exposed, and cell sorting requires a large number of cells to calibrate the machine. Thus, the following issue arises: "Can data obtained from cell populations support hypotheses about single cells?" (Fagan 2013a, 54). Philosopher Melinda Fagan analyzed the structure of this inference. She showed that the conclusion that the observed characters could be attributed to the cells of the isolated subpopulations relies on the following postulate: the isolated cell populations are homogeneous with respect to the studied character, in this case, stemness. However, there are many reasons to doubt that this is the case—i.e., that populations of cells, even stem cells, are homogeneous. Factors such as variability of gene expression, quantity of proteins, cell morphology, and cell-cell interactions are inevitable sources of variation that cast doubt on the ability to infer cellular properties from population-level experiments (Fagan 2013a, 58).

CONCLUSION

Beginning in the 1950s, hematologists pursued investigations within three overlapping research domains. Researchers working on the question of the clonal or multicellular origin of cancers demonstrated the common origin of different cancer lineages—leading to the conclusion that hematopoietic and leukemic stem cells exist. Meanwhile, researchers working on understanding the rate of cancer cell division developed two hypotheses involving stem cells—one in which cancer cells were a self-maintaining system, with some leukemic blast cells being stem cells, and one in which reservoirs of undetected stem cells continually provided cancers with new cells. Finally, within the context of bone marrow transplantation, researchers developed experimental protocols that uncovered the clonal capabilities (differentiation and self-renewal) of hematopoietic cells. It is through the combination of these three research domains that the modern conception of the CSC was derived. By the 1990s, the production of monoclonal antibodies, the commercialization of FACS technology, and the production of immunodeficient mouse strains allowed biologists to distinguish the populations enriched with hematopoietic and leukemic stem cells based on phenotypes and to test their self-renewal and differentiation abilities. Thus, the concepts of "leukemic stem cell" and "hematopoietic stem cell" were established. The existence of stem cells in leukemia was eventually extended to many cancers, which resulted in the framing of the more general concept of the "cancer stem cell" described in Chapter 2.

The analysis of the historical roots of the CSC theory in hematology leads again to the question encountered in Chapter 3—namely, to what cancer cells do the concept of CSC refer? The same question also applies to the more general concept of the stem cell. In addition to this problem, the techniques used to identify and characterize stem cells and CSCs raise conceptual problems about the nature of stemness: do the properties of stemness (self-renewal and differentiation) reside within the individual cells (i.e., they are intrinsic)? Or are they collective properties (i.e., conferred upon an individual by its membership within a population or belonging to an entire population)? These questions will be addressed in Part III.

Together, Chapters 3 and 4 show that the notion of stem cell (either normal or cancerous) has been conceptualized in two ways:

1. The notion of stem cell can refer to a cell that occupies a particular position in the genealogy of a cell lineage. The stem cell is then the cell from which all other cells of a given lineage originate. I refer to this notion of the stem cell as the "genealogical meaning of the stem cell."

2. The notion of the stem cell can refer to a category of cells with particular properties: long-term self-renewal and differentiation. I refer to this notion of the stem cell as the "technical meaning of the stem cell."

Contrary to the common idea that these two meanings were historically successive (e.g., Brandt 2012), these two chapters show that the concepts of stem cell and CSC often contemporaneously bear both the genealogical and the technical meaning. Chapter 5 shows that this double meaning is a point of confusion for the CSC theory—are CSCs stem cells in both the technical and genealogical sense?

III

Debates on CSCs and Stem Cells:
What Are They?

5

ORIGIN, STEMNESS, AND STEM CELLS: THE MEANING OF WORDS

The analysis of the historical roots of the CSC theory showed that the CSC theory emerged laden with conceptual ambiguities. First, many theories of the stem cell origin of cancers have coexisted, either successively or simultaneously, revealing that inside the general hypothesis of a stem cell origin of cancer, the word "stem cell" can refer to different cells. Second, stem cells, within these theories, also displayed both a technical and a genealogical meaning—i.e., they were considered a particular cell type characterized by specific developmental abilities (technical), and they were considered the cells from which the other cancer cells originate (genealogical). This chapter discusses the question of whether the concept of CSC should be understood in a technical meaning, a genealogical meaning, or both.

In addition to parsing the technical versus genealogical meaning of the concept of CSC, the hypothesis of a stem cell origin of cancer also carries ambiguities about how to understand the word "origin." Philosopher Thierry Hoquet (2009, 2010) highlighted that the word "origin" can have two distinct meanings. In German, for example, "origin" can be translated to *"Ursprung"* or *"Entstehung."* *Ursprung* denotes the primitive source (i.e., the material from which derivatives are generated), while *Entstehung* denotes the process of origination. The ambiguity surrounding the meaning of the word origin raises a question about the CSC theory that will also be addressed in this chapter: are CSCs the *Ursprung* or *Entstehung* of cancers? In the first case, cancers would start from mutated stem cells. In the second case, cancers' development would be driven by CSCs. By conflating the two meanings of origin, we assume that CSCs come from mutated stem cells, whereas the origin of CSCs is a matter of debate.

TO WHICH CELLS DOES THE CONCEPT OF CSC REFER?

The phrase "cancer stem cell" is controversial within the scientific literature, prompting biologists to suggest three alternative phrases:

- Cancer-initiating cell
- Cancer-propagating cell
- Cancer stem-like cell or stemloid cell

The pluralist nomenclature that has been developed for CSCs shows that this concept can refer to multiple, and possibly very different, cells.

Cancer-Initiating Cells

Proponents of the expression "cancer-initiating cell" criticize the name "cancer stem cell" for being too abstract and claim that "cancer-initiating cell" reflects more clearly the data and observations obtained in the laboratory (e.g., Singh et al. 2004; Hill and Perris 2007; Zhou et al. 2009a). The phrase "cancer-initiating cell" puts the emphasis on the function of the cell rather than on its identity and gives preference to the genealogical meaning over the technical meaning of stem cell.

It is notable that in the modern history of CSCs, John Dick first called them "SCID Leukemia-Initiating Cells," in reference to their functional ability to initiate leukemia in SCID mice (e.g., Bonnet and

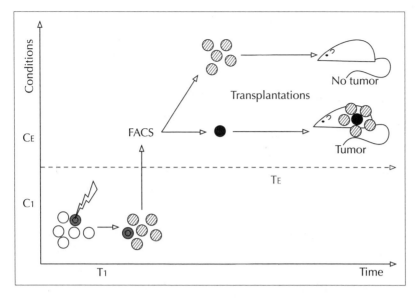

Figure 5.1. To which cells does the concept of CSC refer? The concept of cancer-initiating cells. C_E, experimental conditions; T_E, time during experimentation. White cells are non-cancer cells. Striped cells are non-CSC cancer cells. Gray cells with inner circle are cancer-initiating cells in the patient, and dark gray cells are cancer-initiating cells identified in the laboratory. All gray cells are considered CSCs in the CSC theory. (Illustration © 2016 by Lucie Laplane.)

Dick 1997). It was only a few years later, in 2001, that Tannishta Reya et al. suggested the more abstract name "cancer stem cell," universalizing the concept to all cancers. The suggestion to refer to those cells now called CSCs as "cancer-initiating cells" is a step back from this conceptualization. Endorsing an intermediate position, proponents of the name "cancer-initiating cells" argue that we know what CSCs do (they initiate cancers, not only leukemia but also other types of cancer), but we do not know what they are.

Yet, the word "initiation" is no less ambiguous than are the words "origin" and "stem cell," and the phrase "cancer-initiating cell" can refer to two possibly different populations of cancer cells:

- The cells that initiate cancers under natural conditions (at time T_1 in conditions C_1: see Figure 5.1)
- The cells that initiate cancers in mice during laboratory experiments (at time T_E in conditions C_E: see Figure 5.1)

Laboratory experiments on cancer cells are temporally situated, and there is no proof that the cells that initiate the cancer in a patient (at T_1 in C_1) and those that initiate the cancer in a mouse in a laboratory (at T_E in C_E) are similar. Initiating cells, under laboratory conditions, have been isolated, marked, sorted, and transplanted in a specific biological environment. Such manipulations may affect the so-called cancer-initiating cells (see Figure 5.1).

Several experiments have revealed the impact of the experimental system, used to identify CSCs, on the cells. Model organisms play a major role in the experiments and have significant impacts on the results. NOD-SCID mice, a mouse line developed in the 1990s, were among the first mice in which any cell was able to initiate a tumor (see Chapter 4). Since then, scientists have created new strains of mice into which cancer cells can be grafted. We now know that using different strains of mice can give very different results. In 2008, Sean Morrison's team, at the University of Michigan, used NSG mice, which are NOD-SCID mice that have a supplementary mutation (affecting the interleukin-2 receptor gamma chain) that prevents the maturation of the immune system cells named "natural killer" (NK). Upon transplanting melanoma cancer cells into NSG mice, the team achieved unexpected results: while the ratio of CSCs among cancer cells was estimated to be at around 0.0001 percent in leukemia, in their experiments, Morrison's team found that 25 percent to 27 percent of the cancer cells were able to initiate a melanoma, undermining the idea that CSCs only represent a small fraction of cancer cells (see Chapter 2, hypothesis H_2 of the CSC theory) (Quintana et al. 2008).[1] Similar but less spectacular results had already been obtained for leukemia (Feuring-Buske et al. 2003; Agliano et al. 2008).

Using different mouse strains dramatically changes the population of cells that are identified as being able to initiate a cancer. Even the sex of transplanted mice can affect the result of transplantations. For example, transplantation of cancer cells into female NSG mice can be 11 times more efficient than transplantation into male NSG mice (Notta, Doulatov, and Dick 2010). This raises, once again, the question of the CSC referent.

The fluorescent antibodies used to sort cells can also affect the results of transplantation. Dick's team originally characterized CSCs as being exclusively CD34+CD38-. However, in 2008, a team led by Dominique Bonnet, Dick's former post-doc, at the Cancer Research United

Kingdom London Research Institute, and St. Bartholomew's Hospital, in London, showed that the anti-CD38 antibody that she and Dick used at that time provoked an immune response from the macrophages and the NK cells in the transplanted NOD-SCID mice. This immune reaction results in the elimination of the CD38$^+$ cells carrying the fluorescent antibody. Using a different fluorescent marker, she identified CD34$^+$CD38$^+$ CSCs (Taussig et al. 2008). Two years later, the team's experiments suggested an even greater heterogeneity in the phenotypes of leukemic CSCs (Taussig et al. 2010). In half of the transplantations conducted during these experiments, CSCs were exclusively found in the CD34$^-$ cells and not, as expected, in the CD34$^+$ cells. In the other half, CSCs seemed to be heterogeneous with respect to CD34 and CD38 antigens (some were CD34$^+$CD38$^-$, CD34$^+$CD38$^+$, CD34$^-$CD38$^-$, or even CD34$^-$CD38$^+$). Thus, the fluorescent antibodies used to sort cells before transplantations are not always neutral.

In summary, features of the experimental system like model organism choice and the methods used for cell sorting and transplantations can hugely affect the results of any set of experiments and lead to the identification of different cells as "cancer-initiating cells." Even the length of the experiment (Quintana et al. 2008) and anatomical locations of the transplantations (Yahata et al. 2003; McKenzie et al. 2005) can modify the results of the experiment. Furthermore, the experimental system used to identify CSCs (or cancer-initiating cells) introduces a selection pressure different from what the initial cells experienced when initiating the cancer. Cancer cells that are able to initiate cancers after transplantations in immunodeficient mice are possibly selected during the different experimental phases. It remains to be proven whether the cells that initiated the cancers in the first place are the same cells that are able to initiate cancers in the immunodeficient mice. Consequently, the cells initiating cancers in laboratory experiments should be carefully distinguished from the cells initiating cancers in "natural" contexts (see Figure 5.1).

Cancer-Propagating Cells

The term "cancer-initiating cells" faces the criticism that it hides the potential differences between the cells able to initiate cancers in human bodies and the cells able to initiate cancers in mice following cell sorting and transplantations. To overcome this ambiguity, some biologists, such as Andreas Strasser's team at the Walter and Eliza Hall Institute of Medical Research, Melbourne, Australia, have suggested a distinction between "cancer-initiating cells" and "cancer-propagating cells" (Kelly et al. 2007; see also Valent et al. 2012). According to this distinction, the "cancer-initiating cells" would refer to the cells initiating cancers under "natural" conditions (gray cells with an inner circle in Figure 5.1), whereas, the phrase "cancer-propagating cells" would refer to the cells isolated in the laboratory (dark gray cells in Figure 5.1).

Although this is a valuable distinction, because it overcomes the ambiguity of the notion of "cancer-initiating cell," it is also a source of additional confusion. The notion of initiation is standardly coupled with the notion of propagation in oncology: initiation refers to the precancerous mutations and propagation to the final transformation of precancerous cells to a cancerous state. This leads to ambivalence in the use of the phrase "cancer-propagating cells." For instance, Tariq Enver and Mel Greaves's research group uncovered mutations in hematopoietic stem cells upstream of the actual cancerous transformations. Following this observation, they suggested to distinguish between "cancer-initiating cells" as the precancerous mutated cells and "cancer-propagating cells" as the cancerous cells (Hong et al. 2008; see also Clarke et al. 2006; Curtis et al. 2010). Thus, the cells that Enver and Greaves call "cancer-propagating cells" are instead referred to as "cancer-initiating cells" by Kelly et al. and Valent et al., who use the name "cancer-propagating cells" to refer to the cells that graft in mice.

Jane Visvader, a specialist in breast cancers and normal breast stem cells, provides an additional example of Enver and Greaves's distinction between "cancer-initiating cells" and "cancer-propagating cells":

> It is important to note that the cell of origin, the normal cell that acquires the first cancer-promoting mutation(s), is not necessarily related to the cancer stem cell (CSC), the cellular subset within the tumour that uniquely sustains malignant growth. That is, the cell-of-origin and CSC concepts refer to cancer-initiating cells and cancer-propagating cells, respectively. (Visvader 2011, 314)

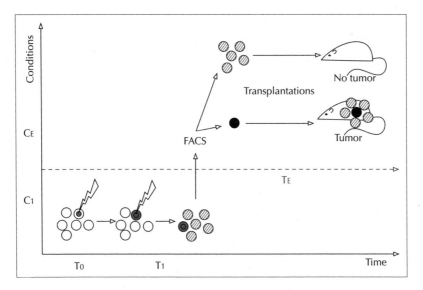

Figure 5.2. To which cells does the concept of CSC refer? The concept of cancer-propagating cells. C_E, experimental conditions; T_E, time during experimentation. White cells are non-cancer cells. Striped cells are non-CSC cancer cells. Dark gray cells are "cancer-propagating cells" identified in the laboratory according to Kelly et al. (2007). Gray cells with an inner circle are "cancer-initiating cells" according to Kelly et al. (2007) and "cancer-propagating cells" according to Hong et al. (2008). White cells with a gray inner circle are "cancer-initiating cells" according to Hong et al. (2008) and Visvader (2011). All gray cells are considered CSCs in the CSC theory. (Illustration © 2016 by Lucie Laplane.)

Visvader's quotation underscores the ambiguities that this chapter seeks to bring to light:

1. To which cells does the concept of CSC refer?
2. To what does the term "origin" refer when we say that CSCs are the origin of cancers?

The names "cancer-initiating cells" and "cancer-propagating cells" highlight that the concept of CSC can refer to at least three potentially different populations of cells, with no consensus on how to name them:

- Precancerous mutated cells, called "cancer-initiating cells" by Greaves and Visvader (the white cell with a gray inner circle in T_0 C_1 in Figure 5.2)
- Cancerous cells that initiate cancers in patients, called "cancer-initiating cells" by Kelly et al. and Valent et al. and

"cancer-propagating cells" by Greaves and Visvader (gray cells with inner circle in $T_1 C_1$ in Figure 5.2)

- Cancerous cells that initiate cancers when transplanted in mice, called "cancer-propagating cells" by Kelly et al. and Valent et al. (dark gray cells in $T_E C_E$ in Figure 5.2)

Evolution and Metastasis

The conditions in which cancer cells develop are in constant flux, giving CSCs the opportunity to evolve over time within a given cancer. Cancer cells are subject to accumulating mutations as well as losses and gains of portions of chromosomes, respectively referred to as genetic and chromosomal instabilities (Lengauer, Kinzler, and Vogelstein 1998; Stratton 2011).[2] Genetic and chromosomal instabilities are ubiquitous in cancers. On the basis of this observation, Peter Nowell (1976) and John Cairns (1975), among others, formulated the clonal evolution theory, according to which "acquired genetic liability permits stepwise selection of variant sublines and underlies tumor progression" (Nowell 1976, 23).[3]

While the CSC theory was formulated in opposition to the stochastic model and clonal evolution theory of carcinogenesis, this opposition is subject to further inspection. At the theoretical level, the CSC theory conflicts with the clonal evolution theory for explanations and predictions of cancer development. The imperative, then, is to know which of these theories gives the best set of explanations and predictions for different cancers. We have already seen that the CSC theory explains and predicts more phenomena using fewer hypotheses than its competitors (see Chapter 2). Furthermore, philosopher Pierre-Luc Germain (2012) has raised issues about the status of adaptive evolution of cancer cells. He highlighted that cancer cells are not paradigmatic "Darwinian populations" (Godfrey-Smith 2009)—that is, evolutionary processes have a lesser role in the explanation of cancers' development. However, new cancer cell lineages are created through time, invoking evolution on a descriptive level. Therefore, at a descriptive level, the clonal evolution and CSC theories are not mutually exclusive (Polyak 2007).

This raises the question of the impact of evolution on CSCs. Do CSC populations evolve? Data suggest that CSC populations themselves are prone to evolve, provoking a growing number of researchers in recent years to suggest integrating the clonal evolution theory into the CSC theory. John Dick's team first highlighted evolution in CSCs in human acute leukemia over the course of serial transplantations in mice (Barabe et al. 2007). They then demonstrated the appearance of genetic diversity among CSCs in BCR-ABL1 lymphoblastic leukemias (Notta et al. 2011) and colorectal cancers (Kreso et al. 2013). Leda Dalprà's and Angelo Luigi Vescovi's team, in Milan, have also highlighted the coexistence of two populations of CSCs with functional and phenotypic differences in glioblastoma (Piccirillo et al. 2009). Finally, studying clonal evolution, Mel Greaves's team demonstrated the presence of CSCs in different subclones in childhood acute lymphoblastic leukemia (Anderson et al. 2011) and glioblastoma (Piccirillo et al. 2015).

According to some biologists, such as Sharmila Bapat and Eric Lagasse, the evolution of CSCs could lead to the emergence of CSC populations with novel functional properties, like the ability to initiate secondary cancers. According to this hypothesis, "not every CSC would be capable of metastasis" (Odoux et al. 2008, 6940; see also Bapat 2007). If this is true, it would be necessary to distinguish a new population of CSCs: metastatic CSCs. Thomas Brabletz and colleagues suggested calling them mCSCs for "metastatic CSCs." According to them, mCSCs emerge additively: mCSCs possess CSC properties plus a new property—mobility (Brabletz et al. 2005; see also Takebe and Ivy 2010). According to Gabriel Ghiaur, Jonathan Gerber, and Richard Jones, the experimental context in which CSCs are usually studied would be a better model for the study of mCSCs than for precancerous CSCs or primary cancer-initiating CSCs because, to initiate a new cancer during transplantation in mice, the cancer cells must be mobile, as the mCSCs are (Ghiaur, Gerber, and Jones 2012).

Thus, time and environmental conditions can introduce diachronic and/or synchronic variations among CSCs, resulting in multiple subpopulations of CSCs, which possibly possess different properties (see Figure 5.3).

Conclusions on the Heterogeneity among CSCs

The alternative terms that researchers have derived for CSCs (cancer-initiating cells, cancer-propagating cells, and cancer metastatic cells),

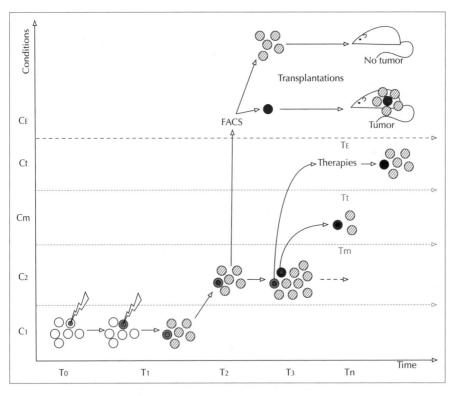

Figure 5.3. To which cells does the concept of CSC refer? Evolution of CSCs, metastasis, and therapies. C_E, experimental conditions; Cm, metastatic environmental conditions (new tissue or organ); Ct, environmental conditions during and after therapies; T_E, time during experimentation; Tm, time during metastasis; Tt, time during and after therapies. White cells are non-cancer cells. Striped cells are non-CSC cancer cells. Dark gray cells in C_E are "cancer-propagating cells" identified in the laboratory according to Kelly et al. (2007). Gray cells with an inner circle in C_1 are "cancer-initiating cells" according to Kelly et al. (2007) and "cancer-propagating cells" according to Hong et al. (2008). White cells with a gray inner circle in C_1 are "cancer-initiating cells" according to Hong et al. (2008) and Visvader (2011). Light to dark gray cells with an inner circle in C_2, Cm, and Ct are CSCs with genetic and/or epigenetic modifications due to clonal evolution in the tumor (C_2), in metastases (Cm), and following therapies (Tt). All gray cells are considered CSCs in the CSC theory. (Illustration © 2016 by Lucie Laplane.)

while useful in a variety of ways, avoid any reference to stemness, in the technical meaning. This creates a theoretical problem. *Every* cancer is *initiated* and *propagated* by some cells. This remains necessarily true, whether the CSC theory is true or false. Therefore, contrary to John Lee and Meenhard Herlyn's statement that "'tumor-initiating cell' has now become synonymous with cancer stem cell" (Lee and Herlyn 2007, 603), and as noticed by Jane Visvader and Geoffrey Lindeman,

> The term "tumor-initiating" is generally used by the field as an operational term to define cells that initiate tumors upon transplantation but it is not necessarily synonymous with a CSC. Tumors that do not follow a CSC model also contain tumor-initiating cells but these do not exhibit stem cell-like properties. (Visvader and Lindeman 2012, 718–719)

My analysis of the denominations "cancer-initiating cells" and "cancer-propagating cells" highlight three features of the concept of CSC:

1. The concept of CSC might hide heterogeneous cells.
2. CSCs (as heterogeneous as they can be) are stem cells in the genealogical sense, that is, they are the cells from which the other cancer cells come.
3. The concept of CSC has more meanings than just the genealogical meaning of stem cell because it implies that only *some* cells will be cancer stem cells.

ARE CSCs TRANSFORMED STEM CELLS?

During the 2006 AACR international congress on CSCs, one of the major issues at hand was the name "cancer stem cell." Michael Clarke and colleagues worried that this name could be misleading regarding the origin of CSCs:

> The term cancer stem cell has led to some confusion. Many interpret the term cancer stem cell to mean that such cells derive from the stem cells of the corresponding tissue. Cancer stem cells may indeed arise from normal stem cells by mutation of genes that make the stem cells cancerous, but this may not be the case in all tumors. (Clarke et al. 2006, 9340)

Shi-Ming Tu, a medical oncologist affiliated with the University of Texas MD Anderson Cancer Center, in Houston, Texas, even consider the hypothesis of "a stem-cell origin of cancer to be a myth" (Tu 2010,

8). This ambiguity around the name "cancer stem cell" led some biologists to use the names "cancer stem-like cells" (e.g., Maenhaut et al. 2010; Gibbs et al. 2005; Al-Assar et al. 2009) and "stemloid cells" instead (e.g., Blagosklonny 2007; Albini, Cesana, and Noonan 2011).

As early as 1995, the French team headed by Bruno Varet had shown that a subtype of acute myeloid leukemia originates from the transformation of non-stem cells (Turhan et al. 1995). However, the doubt about the origin of CSCs from stem cells came, above all, from attempts to model leukemia in mice. With the appearance of new technologies allowing scientists to introduce new genes into cells (transgenesis) or to inactivate existing genes (knockout), along with the accumulation of knowledge on some genetic mutations involved in cancers, biologists became able to produce new mouse models for human diseases, including leukemia. To mimic myeloid leukemia, biologists induced mutations in cells already committed to myeloid differentiation. The emergence of myeloid leukemia following those mutations highlights that cancers can originate from committed progenitors (Cozzio et al. 2003; Passegue et al. 2003; Huntly et al. 2004; see also Gerby 2010, 60–62). Therefore, the origin of cancers from stem cells is not a necessity. Furthermore, Gary Gilliland's team showed, in 2004, that "some, but not all, leukemia oncogenes can confer properties of leukemic stem cells to hematopoietic progenitors destined to undergo apoptotic cell death" (Huntly et al. 2004, 587).

Between 2003 and 2008, a debate arose among CSCs' biologists over the question of the origin of CSCs. Some examples of arguments in favor of the origin of CSCs from stem cells are as follows:

1. *The stem cells' lifetime:* stem cells are the cells that have the longest lifetime in multicellular organisms (see Lansdorp 1997). Cancers result from the accumulation of mutations in a cell over time. Therefore, the longer the lifetime of a cell, the higher the risk to accumulate all the mutations necessary to produce a cancer (e.g., Pardal, Clarke, and Morrison 2003; Dalerba et al. 2007).

2. *The stem cells' self-renewal ability:* stem cells are already able to self-renew. They already have one of the major characteristics of CSCs. Therefore, the hypothesis of the origin of CSCs from stem cells is more parsimonious regarding mutations than the hypothesis of the origin of cancers in non-stem cells. Origin of can-

cers in non-stem cells would require the reactivation of the self-replication machinery (e.g., Passegue et al. 2003).

3. *The similarity between stem cells and CSCs:* normal and cancer stem cells express some common surface markers, such as CD34$^+$CD38$^-$ for hematopoietic and leukemic stem cells. Both also have a similar amount of telomerase (an enzyme that adds DNA sequences at the extremities of chromosomes and is important for long-term self-renewal) (e.g., Huntly and Gilliland 2005).

Max Wicha and his colleagues at the University of Michigan Comprehensive Cancer Center also developed another kind of argument. In Japan, women exposed as teenagers to radiation from the atomic bombs dropped on Hiroshima and Nagasaki developed breast cancers at extremely high rates, sometime years after the exposure. As adolescence is considered the period of breast development where stem cells are present in large number, they hypothesized that stem cells must be targets of cancerous transformations (Wicha, Liu, and Dontu 2006).

Major arguments in favor of the hypothesis that CSCs come from non-stem cells are as follows:

1. *Mouse models of leukemia:* experiments have highlighted that cancerous mutations can affect non-stem cells and result in cancers that are similar to those derived from stem cells (e.g., Passegue et al. 2003; Gerby 2010).

2. *Low numbers of stem cells:* stem cells comprise a tiny subpopulation of cells in tissues and organs. Therefore, they represent a very small target for cancerous mutations. The probability that stem cells become CSCs does not match the probability for a cancer to develop (e.g., Rapp, Ceteci, and Schreck 2008; Chaffer and Weinberg 2015).

3. *Stem cell adaptation:* stem cells exhibit more efficient defense mechanisms against mutations than non-stem cells. This phenomenon, classically explained as an adaptation, implies that the cancerous transformation of stem cells is less probable than the transformation of non-stem cells (e.g., Huntly and Gilliland 2005; Wang and Dick 2005).

4. *Accumulation of mutations during cell division:* the initiation-propagation model (explained above) and, more generally, the

idea that cancers result from an accumulation of mutations make plausible the idea that a first mutation might happen in a stem cell and the following ones, which result in the production of CSCs, only occur in its daughter cells (Polyak and Hahn 2006).

While the history of CSCs given in Part II showed that the CSC theory emerged from a conflated notion of "origin"—in every theory of the stem cell origin of cancers, stem cells (or their equivalents) were considered both the *Ursprung* and the *Entstehung* of cancers—the debates over whether or not CSCs are mutated stem cells ended with the consensus that the CSC theory stands independent from the question of whether CSCs originate from stem cells or non-stem cells, that is, the CSC theory is mainly a theory of the *Entstehung* of cancers. It is possible and even probable that some CSCs come from stem cells, whereas others come from non-stem cells. Regardless, the origin of CSCs affects neither the definition of CSCs nor the functional role they have in the CSC theory (Clarke et al. 2006; Dick 2008; Vermeulen et al. 2008). However, the possibility that non-stem cells could become CSCs raises questions about what it means to be a stem cell.

CONCLUSION

There is a great deal of ambiguity latent within the concept of CSC. Recently, researchers have debated about how to name CSCs. The variety of new terms that they have suggested to replace "cancer stem cell" shows that the concept of CSC refers to multiple, potentially different, populations of cells. Factors such as time and environment can affect CSCs, resulting in phenotypic and functional differences. These differences are important for oncology because they lead to questions that affect the development of CSC-targeting therapies, such as, which of these cells should be targeted? Can we target all of them at once? These questions will be addressed in Part IV.

The issue of how to define the "origin" of CSCs leads to deeper questions about what constitutes the referent of this concept. Historically, scientists working on CSCs have largely ignored the ambiguity of the origin of stem cells by conflating *Ursprung* and *Entstehung.* But recent data showed that CSCs can actually come from two different sources:

either they come from stem cells that have accumulated mutations and transformed into CSCs, or they come from non-stem cells that have acquired stemness following mutations.

Ultimately, the ambiguity surrounding the concept of CSC that has surfaced with respect to their origins and referents leads to one massive, underlying question: what does it mean to be a stem cell (i.e., what is the stemness identity)? The next chapter shows that the concept of stem cell, which includes CSCs, also faces numerous difficulties.

6

STEM CELL IDENTITY

Historical and contemporaneous studies about CSCs draw our attention to the more general concept of stem cell. To understand CSCs, it is important to understand what kind of stem cells they are. However, this chapter shows that the concept of stem cell is itself subject to many debates that cluster around three problems: (1) how to define stem cells (i.e., can we define stem cells by the ability to self-renew and differentiate?); (2) how to characterize stem cells (i.e., can we find a phenotypic or molecular signature that would be specific to stem cells?); and (3) how to understand stem cells (i.e., is stemness specific to stem cells or can non-stem cells acquire stemness?).

Two properties classically define stem cells: the ability to self-renew and the ability to differentiate. The self-renewal property refers to the ability of a cell to give rise to at least one daughter cell identical to itself; stem cells are characterized by the ability to long-term self-renew. The differentiation property refers to the ability of a cell to give rise to at least one more committed differentiated daughter cell—i.e., one that will become a specific type of differentiated cell. Stem cells are able to give rise to multiple kinds of differentiated cells, a property referred to as multipotency, pluripotency, or totipotency. Multipotent stem cells are able to give rise to all the cell types of their tissue. Pluripotent stem cells are able to produce all the cell types of the organism, but they cannot produce the cells of the extraembryonic tissues (placenta, yolk sac, supporting tissues). Totipotent stem cells can produce all the cell types, including extraembryonic ones.[1]

As no two cells are ever identical or different in every respect, philosopher Melinda Fagan framed a more stringent definition of what biologists mean by self-renewal and differentiation. Sameness and difference of cells, she argues, "is relative to some set of characters, such as size, shape, and concentration of a particular molecule. At a given time each cell has some value for a given character, and it is comparisons among these values that determine sameness or difference in any particular case" (Fagan 2013a, 21). She thus suggests the following definition of self-renewal and differentiation:

> Self-renewal occurs within cell lineage L, relative to a set of characters C, for duration τ, if and only if offspring cells have the same values for those characters as the parent cell(s). (Fagan 2013a, 22)

> Differentiation occurs within cell lineage L, during interval t_1–t_2, if and only if some cells in L change their traits such that (i) cells of L at t_2 vary more with respect to characters C than at t_1 or (ii) cells of L at t_2 have traits more similar to traits $C_{m1} \ldots C_{mk}$ of mature cell types $\{1, \ldots k\}$ than at t_1. (Fagan 2013a, 24)

These definitions clarify the properties of self-renewal and differentiation and provide a clear basis to analyze conceptual difficulties related to stemness. Defining stem cells by the properties of self-renewal and differentiation raises two issues developed hereafter. First, the demonstration of the ability of one single cell to both self-renew and differentiate faces paradoxes, frequently referred to as the stem cell "uncertainty principle," which makes it impossible to prove that individual cells have both properties (Potten and Loeffler 1990; Loeffler and Roeder 2002; Seaberg and van der Kooy 2003; Lawrence 2004; Tu 2010; Fagan 2013b, 2015). Second, such a definition meets with two kinds of criticisms from biologists, who argue that these properties are either too inclusive or too exclusive to distinguish stem cells from non-stem cells.

The first issue that the definition of stem cell faces is the difficult, if not impossible, demonstration of the ability of the cells to both self-renew and differentiate. Experiments conducted to demonstrate self-renewal and differentiation abilities necessitate cell divisions. Therefore, the identification of stemness is always retrospective and a posteriori. To overcome this difficulty, biologists try to identify molecular signatures that would be specific to stem cells (cell surface proteins, genes expression, etc.). But no such signature has yet been identified. Additionally, to demonstrate the ability of a cell to self-renew, one has to demonstrate that after a stem cell divides, at least one of the offspring cells is a stem cell. This means that one has to establish that one of the daughter cells is able to self-renew, creating a vicious circle. Furthermore, testing the self-renewal and the differentiation capacities of cells requires the use of a different culture medium. Therefore, no single experiment can assess both properties for a single cell. However, single-cell transplantation assays, although imperfect, offer clues as to the ability of cells to self-renew and differentiate. For example, if a cell is able to participate in hematopoiesis by producing cells of all the blood cell types over a long period of time, it is reasonable to consider that this cell underwent long-term self-renewal and was at least multipotent. The barcoding technology that consists of tagging single cells with DNA sequences (barcodes) that remain present in cells' descendants might also help to resolve this issue (Perié et al. 2014). If one barcode is found in every hematopoietic cell type over a long period of time, then it is reasonable to conclude that the cell that was initially tagged had both extensive self-renewal and differentiation abilities.

The second issue related to the stem cell definition is whether self-renewal and differentiation are adequate properties to define stem cells.

As these properties can be found in other cells, biologists have questioned the adequacy of such a definition (Lander 2009; Mikkers and Frisen 2005). Some non-stem cells, like macrophages (Sieweke and Allen 2013), effector memory T cells (Younes et al. 2003; Schwendemann et al. 2005), and central memory T cells (Miles et al. 2005; Zhang et al. 2005), are capable of long-term self-renewal. Other cells, like early progenitors, can have the same differentiation potential as the stem cells from which they originate (Back et al. 2004; Muller-Sieburg et al. 2004; Trentin et al. 2004). Thus, separately, self-renewal and differentiation are too inclusive to distinguish stem cells from non-stem cells. Most biologists argue that the individual properties are not specific to stem cells but that the accumulation of both properties is. Self-renewal and differentiation, together, would represent the set of necessary and sufficient properties defining stem cells (Seaberg and van der Kooy 2003). However, Dov Zipori and several other stem cell biologists have disputed this claim, arguing that some stem cells lack one of those two properties. Pluripotent stem cells present in the mammalian embryo, for example, are transient and do not undergo long-term self-renewal. The satellite cells and some germinal stem cells are not multipotent but unipotent, that is, they can only give rise to one cell type (Shostak 2006; Zipori 2009). Given these counterexamples, three reactions are possible: first, that some cells are incorrectly classified as stem cells (or as non-stem cells); second, that self-renewal and differentiation are not the right criteria through which to define stem cells; or third, that there is a continuum from stem cells to non-stem cells with no rigorous boundaries. Endorsing this last option, biologists such as Dov Zipori have questioned the very existence of a natural delimitation between stem cells and non-stem cells, that is, they have questioned the existence of a stem cell natural kind.

Homeostatic Property Cluster

The traditional view of essentialism defines natural kinds by the sharing of a common essence, definable by a set of necessary and sufficient properties (Bird and Tobin 2012). Because self-renewal and differentiation are neither necessary nor sufficient, invoking traditional essentialism leads to the conclusion that the stem cell category is artificial rather than natural. However, the concept of "natural kind," as well as the understanding of essentialism associated with it, is highly debated.[2] Some philosophers have suggested a revised notion of essentialism coupled with the use of a different type of definition. Among them, Richard Boyd framed the homeostatic property cluster (HPC) theory according to which the members of a natural kind share some properties of a stable cluster of properties, rather than a set of necessary and sufficient properties (Boyd 1999a, 1999b). Using the HPC theory, Robert Wilson, Matthew Barker, and Ingo Brigandt (2007) argued that stem cells (as well as genes and species) are natural kinds.

Richard Boyd introduced the concept of "homeostatic property cluster" in the context of debates over whether biological species are natural kinds. Traditionally, biological species, such as tigers, were considered a paradigmatic example of natural kinds. The evolutionary perspective that Charles Darwin introduced into biology put this idea into crisis. If, as Darwin claimed, species gradually evolve through descent with modification—that is, if one species can give rise to another through small, accumulated steps—then how is it possible to define the boundaries between them? If species are always gradually evolving, then how could they be defined by necessary and sufficient properties? David Hull pointed out that the static, Aristotelian essentialism and the definitions by necessary and sufficient properties are inappropriate for biological species, whose members are defined by change (Hull 1965; see also Ereshefsky 2012; Bird and Tobin 2012). Instead, he claimed, "taxa names can be defined only *by sets of statistically covarying properties arranged in indefinitely long disjunctive definitions*" (Hull 1965, 323). In this kind of definition by clusters, no property or set of properties is necessary, and any combination of properties can be sufficient (with the number of combinations being indefinite). The HPC theory of Boyd is a modified version of Hull's suggestion. An HPC definition would also provide a set of covarying properties arranged in indefinitely long disjunctive definitions. But, where Hull rejected essentialism (e.g., Hull 1978), Boyd endorsed a renewed essentialism. Indeed, according to the HPC theory, members of a natural kind share some properties of a cluster, which make them similar, as the result of "homeostatic causal mechanisms." That is to say, entities of a natural kind are similar because they share some causal mechanisms that result in the coexpression of some properties (Boyd 1999a, 1999b; Griffiths 1999; Millikan 1999; Wilson 1999).

Robert Wilson, Matthew Barker, and Ingo Brigandt published a defense of the HPC theory using three examples of what they argue to be natural kinds: genes, species, and stem cells. After acknowledging the heterogeneity among stem cells, they suggested the following HPC definition:

- morphologically undifferentiated
- ability of self-renewal (cell division with at least one daughter cell of the same type) over an extended period of time
- ability to give rise to various differentiated cell types (pluripotency, or at least multipotency)
- developmentally derived from certain cells or tissues
- located in specific parts of tissues
- particular complex profile of gene expression and presence of transcription factors
- found in certain cellular-molecular microenvironment ("niche"), which influences the stem cell's behavior
- low rate of cell division

In spite of these properties being typically correlated and shared by most stems cells (but not by more differentiated cells), there are exceptions to each, so that the above describes a genuine HPC kind. (Wilson et al. 2007, 208)

In accordance with the HPC theory, there are exceptions for each of these properties. None of these properties are required, and different subsets may be sufficient. Thus, stem cells of the mouse embryo may belong to the category of stem cells even though they do not self-renew over a long period of time. The same applies to germ cells despite their unipotency. This HPC theory highlights that, at least at a theoretical level, heterogeneity among stem cells does not inhibit the possibility that stem cells belong to a natural kind.

While potentially fruitful, Wilson et al.'s HPC account of stem cells suffers from two problems, one theoretical and one empirical. At a theoretical level, they offer no arguments to demonstrate the homeostasis of the aforementioned stem cell properties. At an empirical level, this HPC definition is not useful for someone who wants to distinguish stem cells from non-stem cells. How can we know if some identified cells are stem cells or not? How can we know that the multipotent hematopoietic progenitors are not stem cells? After all, they have a great number of the properties listed in the HPC definition of stem cells.

A Binominal Definition of Stem Cells

Stem cell is one of those biological concepts torn between two conflicting requirements: the need for fuzzy definitions to allow communication between a variety of specialists and the need for stringent definitions in specific research activities. This tension has been debated in the case of the concept of "gene." On one hand, the vagueness of the concept of gene has been seen as a factor of fecundity and the search for more precise definitions as an epistemological obstacle (e.g., Rheinberger 2000). On the other, empirical applications, such as genome annotation for the Human Genome Project, require precise definitions referring to a specific object. The concepts of self and non-self in immunology also face this issue. Ilana Löwy (1991) and Eileen Crist and Alfred Tauber (1999) have claimed that the strength of these concepts lies in their vagueness. Conversely, Thomas Pradeu (2012) has argued that this vagueness has impeded substantial conceptual distinctions, at the expense of immunology. The concept of stem cell also instantiates this tension. As a boundary object—an object studied by different scientific communities—its fuzzy definition is valuable. It enables collaborations and dialogues between research communities. But as a medical tool or as a medical target, stem cells must be very precisely and carefully characterized.

In order to provide an account of stem cells that is both theoretically and empirically cogent and tractable, I have developed what I term a "binominal definition" of them, that is, a set of definitions that combine one broad generic definition with multiple specific and precise context-dependent definitions. This binominal definition allows both to define stem cells and to distinguish them from non-stem cells. In this section, I defend three claims:

1. It is possible to combine fuzzy and stringent definitions together in what I call a "binominal definition" and, thereby, to reconcile these two conflicting needs for precision and fuzziness.
2. Binominal definitions can overcome the problem of heterogeneity faced by biological entities due to their past and ongoing evolution.

3. Binominal definitions are compatible with the HPC view and can be used to resolve its empirical shortcomings, that is, they can be used to distinguish stem cells from non-stem cells.

The idea of a binominal definition arose in response to an article on gene definitions by Richard Burian. In this article, Richard Burian distinguished two successive types of definition. One, which he named "generic," leaves open the exact reference of the term. The other, which he named "specific," aggregates several discontinuous definitions of the genes. Specific definitions solve the lack of determined reference problem but create conceptual discontinuities, Richard Burian concluded: "What counts as a gene is thoroughly context-dependent . . . this has the consequence that precise definitions of genes must be abandoned, for there are simply too many kinds of genes, delimited in too many ways" (Burian 2005, 175).

I depart from Burian's conclusion and suggest that, at least in the case of stem cells, there is no contradiction in the coexistence of a "generic" definition of the concept with several "specific" definitions. Instead, a generic definition, such as "stem cells are the cells from which tissues develop and are maintained—that is the cells of origin in both the *Ursprung* and the *Entstehung* meaning of origin," allows the concept to perform its functions as a boundary concept. That is, the generic definition allows for dialogue and cooperation between communities of researchers working on different types of stem cells and in different contexts. Specific definitions, provided they do not conflict with the generic definition, enable scientists to assign specific properties to the subcategories of stem cells. For example, one specific definition would characterize hematopoietic stem cells by long-term self-renewal *in vivo*, multipotency, quiescence, location in a specific niche, and asymmetric division. Another specific definition would characterize intestinal stem cells by long-term self-renewal *in vivo*, multipotency, and asymmetric division. And so on for every subtype of stem cell that biologists acknowledge. These specific definitions allow scientists, in turn, to distinguish stem cells from non-stem cells in specific contexts.

The stem cell binominal definition is compatible with the HPC definition. The set of properties of the HPC definition of stem cells can be rearranged into a binominal definition. In both cases, the subsets of properties that stem cells express vary from one type of stem cell to

Table 6.1. Hierarchical structure of the binominal definition of stem cells (SC).

| | | | SC subtype C | |
	SC subtype A	SC subtype B	Sub-subtype 1	Sub-subtype 2
Generic definition	E.g.: Stem cells are the cells from which tissues develop and maintain.			
Specific definition	Property *a*	Property *a*	Property *b*	Property *a*
	Property *b*	Property *c*	Property *c*	Property *d*
	Property *c*	Property *d*	Property *d*	Property *e*

another (see Table 6.1). However, the binominal definition points out regular distributions of properties among subclasses of stem cells. Heterogeneity among stem cells is mainly an outcome of evolution. The binominal definition can account for the evolution of stem cells in two ways. Evolution has produced diverse subtypes of stem cells, and the binominal definition aims at capturing the specificity of each of them to be able to distinguish stem cells from non-stem cells in particular contexts. Stem cells are also subject to evolution, which means that current subtypes of stem cells can gain or lose properties, creating new subtypes. The binominal definition is an open definition, like the HPC definition by Richard Boyd and the definition by clusters by David Hull. The number of specific definitions is indefinite, and specific definitions are context dependent. Therefore, new specific definitions can be added when new stem cells subtypes are discovered, and specific definitions can be reframed according to the knowledge raised by new experiments.

The stem cell binominal definition faces a possible objection in that it only acknowledges heterogeneity among subtypes of stem cells, whereas heterogeneity in stem cells occurs both within and between subtypes. Indeed, I have already mentioned heterogeneity among human hematopoietic stem cells and among leukemic stem cells in Chapter 5. This objection is overcome through understanding that any binominal definition is hierarchically organized; this allows for infinite levels of hierarchy, such that sub-subtypes and further subtypes can be distinguished, if needed. For example, the binominal definition of stem cells one could re-create from current knowledge would distinguish totipotent, pluripotent, and multipotent stem cells. But it would also distinguish the diverse subtypes of pluripotent stem cells, such as embryonic stem cells and induced pluripotent stem cells, and the subtypes of the multipotent stem cells, such as hematopoietic stem cells, intestinal stem cells, or germinal stem cells. And it would also distinguish such sub-subtypes among species, acknowledging the differences between human and mouse hematopoietic stem cells, for example.

MOLECULAR SIGNATURE

In the early 2000s, three groups independently tried to identify a "stemness signature," i.e., a cluster of genes that would be overexpressed in different kinds of stem cells compared to non-stem cells. This research relied on two assumptions:

A1. Stemness (self-renewal and differentiation) qualitatively distinguishes stem cells from non-stem cells.
A2. Stemness is reducible to a set of molecular properties (gene transcripts, cell surface proteins, etc.).

These assumptions rely on the idea that "because all stem cells share fundamental biological properties, they may share a core set of molecular regulatory pathways" (Ivanova et al. 2002, 601).

A1 faces the critiques highlighted in the previous section. On one hand, self-renewal and differentiation are not specific to stem cells. On the other hand, certain stem cells express only one of the two properties. Thus, even if A2 is correct, a molecular definition of stemness would not allow a distinction between stem cells and non-stem cells. Molecular data tend to confirm this diagnosis at several levels. First, the genes highly expressed in stem cell populations are often expressed in the progenitor daughter cells, as well as in some populations of differentiated cells. Second, no universal stemness "molecular signature" that would characterize the various types of stem cells has been identified so far, and the results of such research have generated controversies about stemness identity.

My own research, conducted at the Institute Jacques Monod, under the direction of Michel Vervoort, offers an example of the problem of the continuum in gene expression (Gazave et al. 2013). The objective of our research was to characterize the molecular and cellular signature of the stem cells involved in the posterior growth of the annelid *Platynereis dumerilii*. The sea worm *P. dumerilii*, as is common in a number of bilaterians, has a pool of stem cells located in the caudal part of the body, called the "posterior growth zone" or "segment addition zone" (SAZ). Posterior growth occurs from this SAZ during both embryonic and adult development, as well as during caudal regeneration (Martin and Kimelman 2009). After demonstrating the existence of an SAZ and its role in the posterior growth of *P. dumerilii*, we conducted a series of whole-mount *in situ* hybridization (WMISH) and fluorescent immunostaining experiments, with the aim of examining the pattern of expression of genes known to be expressed in stem cells.[3] Figure 6.1 shows the pattern of expression of some genes (*Piwi, Vasa, Pl10, Nanos, PufB, SmB, Brat, Myc, Id, Ap2, Gcu, Cdx, Evx,* and *Hox3*) in the posterior part of regenerating worms 10 days after amputation, while Figure 6.2 provides an example of a high-magnification view from a confocal microscope of the putative ectodermal stem cells during postcaudal regeneration posterior elongation, with the pattern of expression of *Nanos*. Together, Figures 6.1 and 6.2 highlight the nonspecificity of molecular markers. In both figures, the progenitors of the developing segments (above the row of stem cells) also express these genes, although with a more diffuse pattern, indicating a continuum in gene expression from stem cells to more differentiated cells.

Research on hematopoietic stem cells (HSCs) has shown that HSCs exhibit cell markers also expressed in different tissues (Akashi et al. 2003), as well as markers of differentiated cells of the hematopoietic system itself, including Thy-1, which is a marker of T lymphocytes, and Mac-1, a marker of macrophages (Morrison et al. 1995). This nonspecificity of genetic markers seems widespread. Dov Zipori's team showed that mesenchymal stem cells express neuronal specific genes (Blondheim et al. 2006) and genes involved in the production of membrane receptors of B and T lymphocytes (Lapter et al. 2007 and Barda-Saad et al. 2002, respectively). Nonspecificity of gene expression would reach its climax in embryonic stem cells that would express almost all of the genome (Efroni et al. 2008).[4] According to the hypothesis developed by Michal Golan-Mashiach and Jean-Eudes Dazard, at the Weizmann Institute of Science, in Rehovot, Israel, the more immature the cells are, the more numerous are the genes they express (Golan-Mashiach et al. 2005; see also Zipori 2004).

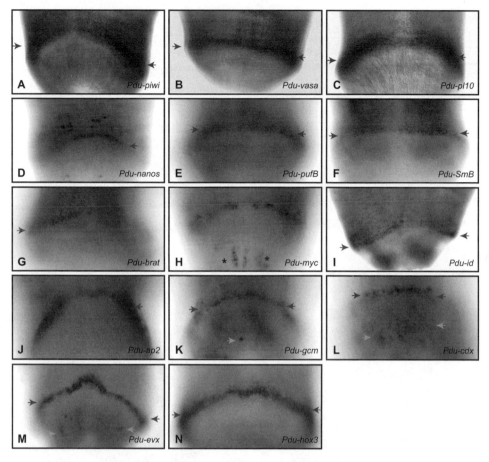

Figure 6.1. Expression of some "stem cell genes" in putative ectodermal stem cells during postcaudal regeneration posterior elongation. Whole-mount *in situ* hybridization for the genes whose name is indicated on each panel is shown. All panels are dorsal views (anterior is up) and only the posteriormost part of the worms is shown. The arrows point to a ring of ectodermal expression immediately anterior to the pygidium, and the asterisks point to an expression in anal cirri. Diffuse staining in nascent segments just in front of the putative ectodermal stem cells is also observed for several genes. (Reprinted with permission from Gazave et al. 2013, Figure 7. © 2013 Elsevier Inc. Reprinted with permission from Elsevier.)

The fact that one or more genes expressed by stem cells are also expressed by non-stem cells does not contradict the possible existence of a complex molecular signature specific to stem cells. The specificity of the stem cells could be found in the expression of a cluster of multiple genes, in which each gene may be independently expressed by non-stem

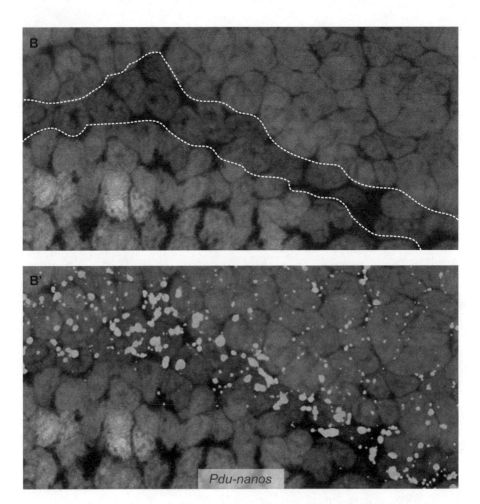

Figure 6.2. High-magnification views of the putative ectodermal stem cells during postcaudal regeneration posterior elongation of *P. dumerilii* (dorsal views of a small posterior part of the worms, anterior up). (B) The dotted lines delineate a row of cells that appear distinct from the more anterior (segment anlagen) and posterior cells (pygidial ectoderm) based on Hoechst labeling. These cells correspond to the posteriormost cells expressing *Pdu-piwi*, *Pdu-nanos*, *Pdu-hox3*, and *Pdu-ap2* (see the original figure for *Pdu-piwi*, *Pdu-hox3*, and *Pdu-ap2*). (B′) Hoechst labeling and fluorescent whole-mount *in situ* hybridization of *Pdu-nanos* (little light dots). (Reprinted with permission from Gazave et al. 2013, Figure 8. © 2013 Elsevier Inc. Reprinted with permission from Elsevier.)

cells. Three leading groups in stem cell research embarked on the search for such a complex molecular signature in the early 2000s. Ihor Lemischka's group at Princeton compared the genetic profiles of human and mouse hematopoietic stem cells and found 283 shared highly expressed genes. They considered some of these genes constitutive of the stemness "genetic program" (Ivanova et al. 2002, 604). Douglas Melton's group at Harvard compared transcription profiles of embryonic, neural, and hematopoietic mouse stem cells and found a list of 216 highly expressed genes (Ramalho-Santos et al. 2002). Finally, Bing Lim's team from the Genome Institute of Singapore compared gene expression profiling of embryonic, neural, and retinal stem cells. They identified 385 common genes (Fortunel et al. 2003). The latter group compared their data with the two previous groups and found only one common gene: *integrin-alpha-6 (ITGA6)*. The *ITGA6* gene codes for the α6 subunit of the α6β4 transmembrane protein, which is by no means specific to stem cells. Indeed, it is primarily found in epithelial differentiated cells.[5]

A debate followed these results, which brought about two major interpretations of the failure to identify a molecular signature of stemness. Most biologists attribute the failure to experimental limits (Burns and Zon 2002; Ivanova et al. 2003; Seaberg and van der Kooy 2003; Vogel 2003). In their views, scientists should work harder to turn the "current impressionistic portrait of a stem cell into a more realistic one" (Burns and Zon 2002, 613). But, for some, the results of these studies constituted a challenge to the very existence of a stemness molecular signature (e.g., Evsikov and Solter 2003). These dissenters concluded that stemness might be slightly different from what is expected, suggesting that maybe "there is no such thing as intrinsic stemness at the molecular level, such that perhaps stemness should be understood as a relational property between cells and their microenvironment generating the functionality of stem cells" (Robert 2004, 1007). In this interpretation, the inability to identify a common molecular signature of stem cells is not attributable to the experimental conditions but results from the fact that stemness is not the essential property (or cluster of essential homeostatic properties) of a natural kind to which all stem cells belong.

Additionally, even if stemness were specific to stem cells and reducible to a molecular signature, Peter Quesenberry and others argue that such a signature would not be sufficient to identify all the stem cells of a population. Indeed, self-renewal and differentiation genes are expressed only by dividing stem cells. But stem cells are often quiescent and only seldom divide; thus, quiescent stem cells do not express the stemness signature and, consequently, would not be identified as stem cells by such a characterization (Quesenberry, Colvin, and Lambert 2002, Colvin, Quesenberry, and Dooner 2006, Quesenberry et al. 2007).

CHALLENGES TO THE STEM CELL DIFFERENTIATION HIERARCHY

In the traditional hierarchical model of stem cells, the stem cell natural kind is at the top of a hierarchy of a gradual differentiation that gives rise to different types of cells (which also represent natural kinds) (see Figure 6.3). This model carries two dogmas:

D1. Differentiation is a one-way street, going from the stem cells to the differentiated cells.
D2. Once engaged in differentiation, cell fate is determined.

These two dogmas are deeply engrained in the minds of biologists. They date back to the germ layer theory, according to which development is irreversible and embryogenesis results in a reduction of the possible fates of the cells (see Chapter 3; see also Maienschein 2014). The following section reviews examples that have led stem cell biologists to question the traditional hierarchical model of stem cells and the two dogmas upon which it rests.

Cloning: Differentiation of the Nucleus Is Reversible

Cloning was the first cell manipulation to cast doubt on the traditional hierarchy of stem cells, indicating that at the level of cell nuclei, differentiation is not inexorable. Cloning is a process of asexual reproduction by nuclear transplantation: the nucleus of a differentiated cell is transplanted into an oocyte from which the nucleus has been removed (an enucleated oocyte). The capacity of the clone to produce a zygote, and then an adult organism, illustrates that a nucleus that has reached the end of the developmental process of differentiation retains a surprising

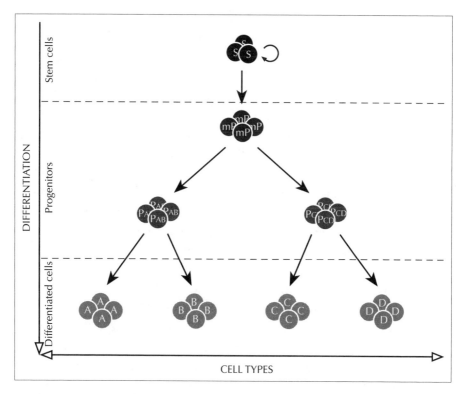

Figure 6.3. The classical differentiation hierarchy model. Stem cells (S) are able to self-renew and to give rise to different cell types (A, B, C, and D), toward the generation of multipotent progenitors (mP) that themselves differentiate into more committed progenitors (P_{AB}, progenitor engaged toward lineages A and B; P_{CD}, progenitors engaged toward lineages C and D). Arrows show irreversibility of differentiation. (Illustration © 2016 by Lucie Laplane.)

amount of plasticity. Although the technique is older, it became famous, inside and outside the scientific community, following the birth of Dolly the sheep in 1996 (Wilmut et al. 1997; on the history of cloning, see Maienschein 2003). The low efficiency of the cloning technique has led some biologists to question the reprogramming capacity of the nucleus. They have hypothesized that the few cells with nuclei capable of producing clones would in fact be stem cells—present in small amounts in adult tissues. This argument was notably supported by Irving Weissman, a prominent representative of the classical representation of stem cells as a natural kind and of differentiation as a one-way process: "I propose that neither dedifferentiation nor transdifferentiation in these instances, but rather that stem cells (whether totipotent stem cells, somatic stem cells, or primitive germ cells) in unexpected sites are responsible" (Weissman 2000, 162).[6]

Rudolf Jaenisch and Konrad Hochedlinger, as well as Atsuo Ogura's team, have brought decisive evidence to show that the nuclei of differentiated cells were truly reprogrammable and could give rise to clones by respectively cloning mice from nuclei of mature lymphocytes and natural killer T (NKT) cells. These cells have specific DNA rearrangements that represent indisputable markers of differentiation, providing evidence that cloning with the nuclei of differentiated cells is indeed possible (Hochedlinger and Jaenisch 2002; Inoue et al. 2005). Recent research on cloning also shows that serial cloning without a decrease in efficiency is possible, suggesting a complete dedifferentiation of the nuclei. Sayaka Wakayama et al. (2013) successfully recloned mice over 25 generations without losing cloning efficiency and with no evidence for the accumulation of reprogramming or genomic errors.

It should be noted that, recently, Shoukhrat Mitalipov's team obtained stable lines of human embryonic stem (ES) cells by nuclear transfer (Tachibana et al. 2013). This success reopens old debates about cloning, although at the present time, a defect in the development of the placenta prevents the implantation of the embryo in the endometrium of the uterus, making any further development impossible (Vogel 2013).

Induced Pluripotent Stem Cells (iPS Cells): Differentiated Cells Can Dedifferentiate to Stem Cells in the Laboratory

After cloning technologies, induced pluripotent stem cells (iPS cells) technologies further breached the principle of irrevocable differentiation. iPS cells are pluripotent stem cells produced in the laboratory from differentiated somatic cells (fibroblasts, for example). The press release of the Nobel Prize awarded to Shinya Yamanaka and John Gurdon in 2012 for this technique reveals the extent of the breach: "These groundbreaking discoveries have completely changed our view of the development and cellular specialization. We now understand that the mature cell does not have to be confined forever to its specialized state" (Nobel Prize 2012).

Several protocols are now available to produce iPS cells. In 2006, Kazutoshi Takahashi and Shinya Yamanaka used retroviruses as viral vectors to inject a cocktail of genes *(Oct4, Sox2, c-Myc, Klf4)* in mouse fibroblasts *in vitro*. These four genes encode for transcription factors expressed in embryonic pluripotent stem cells. After a few weeks of *in vitro* culture, a few differentiated fibroblasts (around 1 of 2,000) dedifferentiated and switched to a pluripotent undifferentiated state (Takahashi and Yamanaka 2006; Takahashi et al. 2007; Okita, Ichisaka, and Yamanaka 2007). Since then, techniques have evolved and diversified (Laplane et al. 2015). It is now possible to use several combinations of genes; proto-oncogenes *c-Myc* and *Klf* are not necessary and can be replaced by other genes such as *Nanog* and *Lin28* (Yu et al. 2007; Silva et al. 2008). Other types of viral vectors such as lentiviruses, adenoviruses, or Sendai viruses can also be used. Lentiviruses have the advantage that they can transfect non-dividing cells (Yu et al. 2007). Adenoviruses and Sendai viruses have the advantage that they do not incorporate their genes into the genome of their host cells—this prevents mutagenic insertions and associated cancer risks (Stadtfeld et al. 2008; Zhou and Freed 2009). Moreover, iPS cells can be generated without using any viruses through the use of plasmids (small non-chromosomal DNA molecules that can replicate and be integrated into the cell to generate the required transcription factors) (Okita et al. 2008), proteins (Bru et al. 2008; Zhou et al. 2009a), or chemicals mimicking the transcription factors (Shi et al. 2008).

Despite recent methodological advances, the production of iPS cells remains a mystery: how or why some cells produce iPS cells, while others do not, is largely unknown. The low efficiency of iPS cell technology has been explained through two alternative interpretations known as the "elite" and the "stochastic" models. According to the elite model, only a subset of any given cells is competent for reprogramming, and iPS cells are not generated from differentiated cells but from this subset of cells that are suspected already to be stem cells. In 2010, Mari Dezawa's team first reported the existence of adult pluripotent stem cells, isolated from cultured skin fibroblasts and from bone marrow; they called these cells "multilineage differentiating stress enduring" (Muse) cells (Kuroda et al. 2010). Dezawa's team tested the elite and stochastic model by comparing the efficiency of iPS cell technology on muse fibroblasts and non-muse fibroblasts. In their experiments, they succeeded in generating true iPS cells from the muse cells, whereas non-muse fibroblasts only partially dedifferentiated. Such results highly favor the elite model (Wakao et al. 2011; Wakao, Kitada, and Dezawa 2013).

In contrast, according to the stochastic model, every cell type has the potential to be reprogrammed. Jacob Hanna et al. (2009) framed a mathematical model that accounts for the stochastic model of reprogramming. Several articles have reported successful reprogramming of T cells (Staerk et al. 2010; Brown et al. 2010) and B cells (Hanna et al. 2008; Choi et al. 2011), in accordance with the stochastic model. Others suggested some forms of integration between the two models. Matthias Stadtfeld and Konrad Hochedlinger (2010) defended the idea that the probability of inducing pluripotent stem cells is related to the level of differentiation. They tested the efficiency of reprogramming in mouse hematopoietic cells at different stages of differentiation and showed that it was drastically decreasing with differentiation (Eminli et al. 2009). Filipe Grácio, Joaquim Cabral, and Bruce Tidor (2013) suggested a mathematical model, according to which "unpredictability and variation in reprogramming decreases as the cell progresses along the induction process, and identifiable groups of cells with elite-seeming behavior can come about by a stochastic process."

Despite the elite versus stochastic controversy, iPS cell technology raises doubt about whether differentiation is an irreversible process. For example, stem cell biologist Dov Zipori interprets iPS cells as evidence that differentiated cells retain a fundamental ability to return to the stem cell state. According to him, such results can only be achieved if "cells, even completely mature ones, would be equipped with a 'return to the stem state' machinery" (Zipori 2009, 202–203).

STAP Cells: Can Stress Trigger Dedifferentiation?

On January 30, 2014, *Nature* published two papers (an article and a letter) written by Haruko Obokata et al. (2014a, 2014b) at the RIKEN Center for Developmental Biology in Kobe, Japan, exhibiting a new way to reprogram somatic cells to a pluripotent state, without using either nuclear transfer (cloning) or stem cell transcription factors (iPS cells).

The technique consisted of transiently culturing cells in a low-pH medium to stress the cells. The process was referred to as "stimulus-triggered acquisition of pluripotency" (STAP). Soon after the publication, the two papers faced several attacks, mainly concerning figures and difficulty encountered by other stem cell biologists when trying to reproduce the method (Cyranoski 2014a). The RIKEN Center started an investigation that resulted in the accusation of scientific misconduct toward Haruko Obokata (Cyranoski 2014b). *Nature* eventually retracted both articles in July 2014.[7]

Despite the retraction of both papers, the question about whether stress and/or other external factors can trigger dedifferentiation has been raised. The accusation of scientific misconduct undermines the results of Obokata et al., but that does not mean that STAP cells cannot exist, leaving us with new unresolved questions about stem cell identity.

Regeneration: Cells Can Dedifferentiate In Vivo

Cloning and iPS cell technology, as well as the more controversial STAP method, highlight the possibility of dedifferentiation but only under artificial conditions. One can argue that we cannot draw any conclusions from these results regarding the nature and identity of stem cells *in vivo*. Yet, some data also indicate that dedifferentiation might occur *in vivo* under particular conditions such as regeneration. In plant cells, for example, dedifferentiation is a commonly accepted process:

> It is well established that differentiated plant cells do not lose their developmental potentialities during normal development but retain plasticity, that is, they are capable to dedifferentiate and acquire new fates. (Grafi 2004, 1)

> Plant cells retain their ability to be a truly totipotent stem cell even when they are fully differentiated. (Lohmann 2008, 9)

Regeneration refers to the renewal of a lost biological part, from cell populations to tissues and organs. In 2010, Kenneth Poss distinguished "homeostatic regeneration" from "facultative regeneration" (Poss 2010). The former refers to the natural replacement of cells lost in day-to-day minor damages and by cell death. The latter refers to tissue replacement after substantial trauma, such as amputations or ablations. Thomas

Hunt Morgan had suggested a similar distinction in 1901, between "physiological regeneration" and "restorative regeneration" (Morgan 1901). Every organism is capable of "homeostatic" or "physiological" regeneration, whereas species differ in their ability to perform "facultative" or "restorative" regeneration. Some data suggest that, at least in some cases, "facultative" or "restorative" regeneration requires a first step of dedifferentiation.

Zebrafish, for example, can support a cardiac amputation up to 20 percent. Two studies suggest that after such an amputation, the regeneration process requires a first phase of dedifferentiation of the cardiac muscle cells (cardiomyocytes). They characterize this first step of dedifferentiation by the loss of cell contact and a return to the cell cycle (Jopling, Boue, and Izpisua Belmonte 2010; Kikuchi et al. 2010). Other studies suggest that zebrafish can also regenerate their retina by dedifferentiation of Müller cells, which are glial cells (supporting cells) present in the retina of vertebrates (Bernardos et al. 2007; Ramachandran, Fausett, and Goldman 2010; Wan, Ramachandran, and Goldman 2012).

Studies of limb regeneration in *Urodela* also indicate that there is a dedifferentiation process prior to the establishment of a "blastema" (Nye et al. 2003). The blastema refers to a group of undifferentiated cells from which regeneration of the lost limb occurs. At least three types of differentiated cells have been reported to dedifferentiate during such regeneration events: fibroblasts (Satoh et al. 2008a), keratinocytes (Satoh et al. 2008b), and myotubes (Echeverri, Clarke, and Tanaka 2001).

Even in mammals, Schwann cells (cells in the peripheral nervous system) are associated with a regenerative ability following a dedifferentiation process. This dedifferentiation is characterized by loss of contact with the axon, reexpression of genes associated with Schwann cell progenitors, and return to proliferation (Chen, Yu, and Strickland 2007). Accumulating data also suggest that in mammals, epithelial cells are plastic and can return to a stem cell state during tissue repair and regeneration (Donati and Watt; Tetteh, Farin, and Clevers 2015; see also Krieger and Simons 2015).

While these studies suggest the occurrence of dedifferentiation during the regenerative process, upon careful examination, the concept of dedifferentiation used in this research is rather vague. First, the return to the stem state is not formally demonstrated. Dedifferentiation could simply indicate a "return to a less differentiated stage" without involving a return to the stem state. Chris Jopling et al., for example, define dedifferentiation as follows: "One of the mechanisms associated with natural regeneration is dedifferentiation, which involves a terminally differentiated cell reverting back to a less-differentiated stage from within its own lineage" (Jopling, Boue, and Izpisua Belmonte 2011, 79). Second, dedifferentiation is often defined by a return of differentiated postmitotic cells to the cell cycle. But, return to the cell cycle does not necessarily mean a return to the stem state or even a return to any earlier stage of differentiation. It could as well refer to a new state that would differ from the classical stage of the hierarchical differentiation.

Juan Carlos Izpisua Belmonte's team at the Salk Institute for Biological Studies in La Jolla, California, compared regeneration with reprogramming of iPS cells (Christen et al. 2010). More precisely, they compared the expression profile of factors associated with pluripotency in blastema cells and in embryonic pluripotent stem cells in the *Xenopus* and zebrafish. Their results highlight some differences, particularly in overexpressed genes and in the cell cycle, as well as some functional similarities, particularly in the involvement of two transcription factors, namely Sox2 and pou5f1—better known under the name of Oct4 in iPS cell research area. The expression of *Sox2* and *pou5f1/Oct4*, two of the four genes used for the production of iPS cells, appears necessary for the creation of the blastema and for regeneration to occur. Wei Zhu et al. obtained similar results for limb regeneration in axolotl (Zhu et al. 2012).

Together, these data, although still incomplete, point to the possibility of at least a partial dedifferentiation of some differentiated cells, indicating that differentiation is not necessarily a one-way street *in vivo*. Little information, however, is available on the factors involved in the induction of dedifferentiation in regeneration, leaving some questions open, such as, why can Müller glia cells dedifferentiate in zebrafish but not in mice?

Niche: Can the Microenvironment Induce Dedifferentiation?

Microenvironment and cell niche are concepts that are becoming increasingly popular in stem cell biology. According to the stem cell niche hypothesis, first framed by Ray Schofield in the late 1970s, stem cell properties (or at least some of them) are regulated by the environment in which the stem cells are embedded. This environment, also called the "microenvironment," consists of a structure called a "niche" containing (1) a stem cell, (2) a "supporting cell" to which the stem cell adheres, and (3) all the soluble factors localized near the stem cell. Accumulating data suggest that the niche regulates self-renewal, differentiation, asymmetric division, quiescence, and, more generally, developmental and regenerative capacities of at least some stem cells in some species. However, many questions are still unresolved, particularly concerning the possible differences between stem cell types and between species. The ubiquity of the role of the stem cell niche remains to be determined. Another unresolved question is whether the niche "regulates" or "determines" stemness. I therefore suggest distinguishing two interpretations of the stem cell niche hypothesis:

- A weak interpretation: the niche regulates stem cells activities.
- A strong interpretation: the niche determines stemness, meaning that the niche can induce stemness in a non-stem cell (i.e., it can induce dedifferentiation).

Research on the stem cell niche in the model organisms *Drosophila melanogaster* (fruit fly) and *Caenorhabditis elegans* (nematode) shows that the supporting cells, which are adjacent to stem cells, secrete factors that are required for the maintenance of stem cell identity and to control stem cell self-renewal. Maintenance and division of germinal stem cells in *Drosophila* females are regulated by growth factors such as dpp protein (decapentaplegic protein), expressed by the niche cells called hub cells, which are essential to activate stem cell division and prevent their differentiation (Xie and Spradling 1998). In *Drosophila* males, the hub cells express the *Upd* factor *(unpaired)* that activates the stem cells' JAK-STAT signaling pathway, which is required for self-renewal. Without activation of this signaling pathway, germ stem cells will differentiate (Kiger et al. 2001; Tulina and Matunis 2001).

Fumio Arai, Toshio Suda, and others have highlighted that the niche can also keep stem cells in a quiescent state, i.e., prevent them from dividing (Arai and Suda 2008). In the hematopoietic system, angiopoietin-1 (Ang-1) and thrombopoietin (THPO), produced by mesenchymal and stromal cells constituting the hematopoietic stem cells niche, bind to Tie-2 and Mpl receptors, respectively. This activation of the receptors results in the activation of the expression of both β-catenin (a binding protein) and cyclin-dependent kinase (CDK) repressors, which induce quiescence. It also represses the expression of *Myc* that regulates the production of β-catenin (Arai et al. 2004; Qian et al. 2007; Yoshihara et al. 2007). Abou-Khalil et al. (2009) identified a similar role for angiopoietin in muscle stem cells. Conversely, the expression of Wnt protein by the niche cells induces self-renewal of the stem cells: Wnt protein binds to the receptor complex LRP5 and/or LRP6, inducing, through a molecular cascade, the stabilization of the production of β-catenin that thus accumulates and induces self-renewal (through activation of target genes by β-catenin) (see Reya et al. 2003; Clevers 2006; Nishikawa and Osawa 2007; Suda and Arai 2008; Clevers, Loh, and Nusse 2014).

Adhesion of the stem cell to a support stromal cell or to the basal lamina enables the maintenance of the stem cell in its niche and participates in regulating its activity (Yamashita 2010). Cellular adhesion can orientate the mitotic spindle (the structure that segregates the chromosomes) during cellular division. Orientation of the spindle apparatus determines where the chromosomes segregate and therefore where the cell divides. Cellular adhesion can orientate the mitotic spindle so that one of the two daughter cells stays in the niche, and the other leaves it—this phenomenon has been observed in *Drosophila* ovaries and testes. The hub cells express proteins that play a crucial role in anchoring one pole of the spindle of the stem cells near their hub cells. As a consequence, the stem cell divides asymmetrically. One of the daughter cells stays in the niche and remains under the influence of the activators of the JAK-STAT signaling pathway, which is implicated in self-renewal, and retains the stemness identity. The other daughter cell leaves the niche, is no longer under the influence of the activators of the JAK-STAT signaling pathway, and therefore can engage in differentiation (Deng and Lin 1997; Yamashita, Jones, and Fuller 2003). Xiaoqing Song and Ting Xie have shown that in female *Drosophila,* the loss of the cadherin, necessary for cell adhesion, causes the loss of the germinal stem cells, illustrating the fundamental role that the niche might play in the maintenance of the stemness identity (Song and Xie 2002; Song et al. 2002). The orientation of cell division by the niche has also been observed in mammals, particularly in the epithelium (Seery and Watt 2000; Lechler and Fuchs 2005) and in striated muscles (Kuang et al. 2007).

The niche can also influence stem cells through mechanical factors, such as the elasticity or stiffness of the microenvironment. Research from Dennis Discher's team has shown that mesenchymal stem cells can differentiate differently depending on the culture medium. In soft matrices that mimic the cerebral environment, mesenchymal stem cells differentiate into neurons. In a rigid environment that mimics bone's environment, they differentiate into osteoblasts. In an intermediate matrix similar to muscle, they differentiate into myogenic cells (Engler et al. 2006).

Finally, research on aging has also drawn attention to the niche. One popular hypothesis postulates that stem cells age along with their niche. Experiments conducted in mice have shown that the transplantation of young stem cells into atrophied testes of two-year-old mice failed to regenerate spermatogenesis, indicating the crucial role of the microenvironment in stemness (Zhang et al. 2006). Similar results were obtained for striated muscle satellite cells in rats (Carlson and Faulkner 1989) and mice (Zacks and Sheff 1982). Furthermore, Irina Conboy showed that the activity of satellite cells of aged mice and their regenerative capacity can be restored by reactivating the Notch pathway, "demonstrating that the intrinsic regenerative capacity of aged satellite cells remains intact" (Conboy et al. 2005, 760; see also Conboy et al. 2003). Conboy's work also highlights the role of a larger systemic environment in controlling the activity of muscle stem cells. She established parabiosis between young and old mice, which consists of joining the circulatory systems of two animals to examine the interaction between muscle stem cells and the circulatory system. She also performed *in vitro* exposure of older satellite cells to a young mouse serum and vice versa. Both kinds of experiments showed a modulation of the behavior of the stem cells, indicating the presence of determinant factors in the circulatory system (Conboy et al. 2005; see also Brack et al. 2007).

Together, these data suggest that the properties that define stem cells (self-renewal, differentiation, asymmetric division, quiescence, and, more generally, developmental/regenerative capacity) are at least regulated, if not determined, by the niche. This leads the proponent of the weak interpretation of the niche hypothesis to the understanding that the properties classically attributed to stem cells should actually be attributed to the stem cell-and-niche complex. The strong interpretation of the niche theory pushes the critic one step further in order to claim that the niche not only regulates stemness but can also induce stemness in non-stem cells. An immediate consequence of this line of thought is the need to reevaluate the idea that differentiation is a one-way street.

The strong interpretation of the niche theory dates back to Ray Schofield in the 1970s. When he developed the stem cell niche hypothesis in the late 1970s, he was working on the CFU-S, those hematopoietic cells that can produce spleen colonies in mice that James Till and Ernest McCulloch had discovered a few years earlier (see Chapter 4). At that time, CFU-S were thought to be hematopoietic stem cells, before successive transplantation assays pointed out their limited self-renewal potential. Ray Schofield investigated the different proliferative abilities of CFU-S compared to hematopoietic stem cells, and he framed the niche hypothesis as an explanation of this difference. He suggested that *in vivo,* the hematopoietic stem cells are anchored to a "niche" that holds them and prevents their maturation. Thus, when the stem cell divides, only one daughter cell can be held in the niche and remains a stem cell. The other daughter cell exits the niche and becomes a CFU-S. Once outside of the niche, he hypothesized, differentiation is inexorable. While seemingly similar to the current weak interpretation of the niche hypothesis, Schofield (1978) further claimed that CFU-S could regain their stem cell identity if they entered a vacant niche. In a paper published in 1983, Schofield wrote,

> Stem cell properties do not reside in one specific cell type in the population but, when necessary, cells other than those normally playing the stem cell role, can have stem cell function imposed upon them by the appropriate microenvironment. . . . There are no cells which are *intrinsically* stem cells. (Schofield 1983, 375)

> The feature which determines whether or not a cell acts as a stem cell is the microenvironment. (Schofield 1983, 378)

This research led him to the conclusion that we should refer to a "stem cell system" rather than referring to stem cells, since they do not maintain their stem cell identity outside of that system.

Schofield's hypothesis has three consequences:

1. At the top of the differentiation hierarchy lies a "stem cell system," not just a stem cell.
2. There is no stem cell natural kind since stemness is not an intrinsic or essential property.
3. Differentiation is not a one-way street since non-stem cells can become stem cells if they enter vacant niches.

The strong interpretation of the stem cell niche hypothesis is far less widespread than the weak interpretation. However, some data suggest that the niche can actually induce stemness in non-stem cells. In *Drosophila,* several teams have shown that germ cells committed to differentiation can dedifferentiate and produce functional stem cells upon contact with an appropriate microenvironment (for female *Drosophila,* see Kai and Spradling 2004; for male *Drosophila,* see Brawley and Matunis 2004; Cheng et al. 2008; Sheng, Brawley, and Matunis 2009). Crista Brawley and Erika Matunis (2004) conducted experiments that showed that a *Drosophila* that has no germinal stem cells at time t_1 can regenerate new germinal stem cells at t_2 through migration of cells to the germinal stem cell niche. Pierre Fouchet's team, at the CEA in Fontenay aux Roses, France, has also shown that such dedifferentiation can happen in the testes of mice (Barroca et al. 2009). They interpreted this phenomenon as an argument in favor of "the emerging concept that the stem-cell identity is not restricted in adults to a definite pool of cells that self-renew, but that stemness could be acquired by differentiating progenitors after tissue injury and throughout life" (Barroca et al. 2009, 190).

Transdifferentiation, Metaplasia, and Cellular Plasticity:
Categories of the Differentiation Hierarchy Are Porous

The dogma related to the cell fate (D2: once engaged, the cell fate is determined) is also being challenged because biologists have observed the phenomena of lineage switches and cell fate switches, loosely referred to as "transdifferentiation," "metaplasia," or "cellular plasticity." Here,

I distinguish two kinds of phenomena: first, the transformation of a differentiated cell from one cell type into a differentiated cell of another cell type and, second, the production of a differentiated cell of one tissue by the stem cell belonging to another tissue.

With regard to the first phenomenon, accumulating data suggest that differentiated cells can transform into other cell types: in *C. elegans,* a rectal cell called Y usually transdifferentiates during larval development into a motor neuron cell called PDA (Jarriault, Schwab, and Greenwald 2008), cells of the iris may transdifferentiate into cells of the lens during lens regeneration in amphibians (Grogg et al. 2005), hepatic cells can transdifferentiate into myofibroblasts during chronic inflammation or liver damage (Bachem et al. 1993; Tsukamoto et al. 2006), pancreatic cells can transdifferentiate into hepatocytes (Shen et al. 2003), and cells of the vascular endothelium can transdifferentiate into smooth muscle (Frid, Kale, and Stenmark 2002).

With regard to the second phenomenon, controversial data have also accumulated suggesting that tissue-specific stem cells might have the capacity to differentiate into cells specific to other tissues. Hematopoietic stem cells could thus give rise to brain cells (Brazelton et al. 2000; Mezey et al. 2000; Priller et al. 2001), muscle cells (Ferrari et al. 1998; LaBarge and Blau 2002), myocardial cells (Orlic et al. 2001), or liver cells (Petersen et al. 1999; Lagasse et al. 2000; Theise et al. 2000a, 2000b). Mesenchymal stem cells may differentiate into blood cells, lung cells, liver cells, or intestinal cells (Pittenger et al. 1999; Jiang et al. 2002). Stem cells of the brain (Clarke et al. 2000; Galli et al. 2003), skin (Liang and Bickenbach 2002), muscles (Jackson, Mi, and Goodell 1999), and adipose tissue (Zuk et al. 2001) could also differentiate into cells of other tissues. These data were the subject of much criticism, particularly from proponents of the traditional hierarchical model (Anderson, Gage, and Weissman 2001; Lemischka 2002; Orkin and Zon 2002; Goodell 2003; Wagers and Weissman 2004). The most common criticism is that what has been interpreted as transdifferentiation would actually be the result of cell fusion—a process where two cells combine to form one (e.g., Hawley and Sobieski 2002; Wurmser and Gage 2002). Also debated were the therapeutic implications of stem cell transdifferentiation. Observations of stem cell transdifferentiations raise the hope that adult stem cells could be used for cell therapies and thereby could be substitutes for embryonic stem cells (ESCs): "if adult stem cells can turn into any-thing, why bother with the controversial ESCs?" (Wells 2002, 15).

I distinguished two kinds of transdifferentiation. The first one re-fers to the transformation of one cell type into another. This kind of transdifferentiation does not directly challenge the stem cell category as it only concerns different types of differentiated cells. The second kind of transdifferentiation concerns stem cells, as it refers to their ability to give rise to unexpected types of differentiated cells. However, it does not challenge the stem cell category. Quite the contrary, stem cell plasticity can even be an argument to highlight the similarity be-tween all stem cells. Therefore, not all the challenges to the hierarchical model of differentiation are challenges to the existence of a stem cell natural kind. However, the possible porosity of the hierarchical model's categories has led some biologists to reinterpret these categories as cel-lular states actively maintained by the organism, rather than as sepa-rated classes of entities with particular intrinsic identities (e.g., Theise and Wilmut 2003). In addition, some phenomena of transdifferentia-tion could begin with dedifferentiation. And this possibility has led some researchers, especially Carol and Leonard Eisenberg, to question the existence of a natural, discrete class of stem cells: if unipotent pro-genitors can be redirected to multiple cell fates, then what distinguishes them from multipotent stem cells? "Although rates of differentiation may exceed that of dedifferentiation and transdifferentiation between phenotypes, thereby providing directionality to lineage acquisition, the implication of a transdifferentiation model of cell diversification is that totality of cell phenotype within an organism is a part of a con-tinuum" (Eisenberg and Eisenberg 2003, 216). All these phenomena (cloning, induced pluripotent stem cells, regeneration, role of niche, and transdifferentiations/metaplasia/plasticity) have also led Dov Zi-pori to a similar position:

> It should have become obvious, at this point, that informa-tion available to date on cellular systems clearly indicates that cells possess a high degree of plasticity. The position of the cell in the differentiation cascade is therefore unstable. It is, however, far more stable when cells assume maturity, at the end of the differentiation process. Nevertheless, even at this

very last stage, the genome does maintain complete capability to turn into a plastic state. . . . Sharp boundaries between differentiation stages cannot be observed. (Zipori 2009, 200)

From the absence of clear boundaries between the cells at different stages of differentiation, Zipori (2009, 201) concluded that "stemness does not describe a cell type, but rather a transient state." Furthermore, Blau, Brazelton, and Weimann (2001, 836) asked, "Is there a universal stem cell?" and Harald Mikkers and Jonas Frisen (2005, 2718) wondered about stem cells: are they the result of "nature or *nurture*"? Thus, the plasticity of the stem cell category and hierarchy has led many to question the nature of stem cells.

CONCLUSION

The idea that stem cells belong to a natural kind has been consistently challenged by stem cell biologists since the 1990s for a number of reasons. First, no set of necessary and sufficient properties can currently define stem cells, and there is no evidence of a molecular signature for stemness. Furthermore, cloning and iPS cell technologies show that differentiation is reversible, undermining the traditional hierarchical conception of stem cells. The traditional hierarchy is further brought into question by the realization that dedifferentiation can occur naturally in some contexts, such as regeneration. Additionally, the stem cell niche (1) might control stemness and (2) may even induce stemness in non-stem cells. Finally, cells can transdifferentiate, highlighting a versatility and porosity in cell identity and in the process of differentiation.

These debates show that the identity of stem cells is uncertain, which leads to the following question: how can we draw a clear picture of the stem cell identity? Part IV of this book proposes a classification that brings order and clarification to the identity of stem cells and therefore to CSCs. A direct benefit of this classification is that it helps us to assess the therapeutic options available to the CSC theory.

IV

The Identity of Stemness and Its
Consequences for Cancer Therapies

7

IF STEMNESS IS A CATEGORICAL
OR A DISPOSITIONAL PROPERTY,
HOW CAN WE CURE CANCERS?

The chapters in Part III highlighted that the concepts of stem cell and CSC are ambiguous and subject to numerous debates. The chapters in Part IV show that the diverse conceptions of stem cells that are present in the literature can be sorted into four views, each of which entails a different stem cell identity. This categorization of stem cell identity has deep implications for the development of therapies against cancers: the ability of therapeutic strategies to cure cancers depends on the kind of property stemness really is. Thus, this chapter and Chapter 8 share a single purpose: to analyze stem cell identity to elucidate the consequences for cancer therapies.

Philosophers have considered the debates about stem cells discussed in Chapter 6 as a conflict between two views: the "entity view" versus the "state view" (Robert 2004; Brown, Kraft, and Martin 2006; Leychkis, Munzer, and Richardson 2009; Kraft 2011; Fagan 2013a). Their cue for this opposition arose in the work of biologist Dov Zipori, whose 2004 article, titled "The Nature of Stem Cells: State Rather Than Entity," depicts the "entity view" as the classical view, according to which stem cells are stable entities with stable intrinsic properties (self-renewal and differentiation). In contrast to the entity view, he suggested a "state view," according to which stemness is an extrinsic property that characterizes a cellular state that most cells can return to (see also Zipori 2009). Other biologists have framed similar alternatives. For example, Helen Blau et al. (2001) distinguished the understanding of the concept of stem cell as describing an "entity" or a "function," and Harald Mikkers and Jonas Frisen (2005) framed a conflict between "nature" and "nurture." Asking whether stem cells belong to a natural kind, as I did in Chapter 6, is still another way to describe this same alternative: "entity" and "nature" refer to the belief that stem cells belong to a natural kind, whereas "state," "function," and "nurture" question this belief. This dichotomy merits further inspection. In this chapter and Chapter 8, I show that both sides of the alternative can and should be subdivided. This chapter focuses on the entity/natural kind view and distinguishes between two possibilities: either stemness is a categorical property, intrinsic and specific to stem cells, or it is a dispositional property, specific to stem cells but whose expression depends on extrinsic stimuli. In each case, I discuss the difficulties that CSC-targeting strategies face and the therapeutic alternatives.

STEMNESS: CATEGORICAL PROPERTY VERSUS DISPOSITION

The traditional conception of stem cells considers them to be discrete, natural entities (i.e., stem cells belong to a natural kind) that share some intrinsic fundamental properties united under the general "stemness" property. In metaphysics, these kinds of properties are called "categorical properties." Categorical properties are properties that are intrinsic and essential to some objects (or classes of objects). For example, the atomic structure of oxygen (eight protons and eight electrons) is a categorical property of oxygen. This atomic structure is intrinsic and essential to oxygen; it explains oxygen's chemical and physical properties. If stemness is a categorical property, then stemness exists in virtue of what stem cells are, without references to any kind of interactions stem cells can have with the world.

As we shall see, the categorical conception of stemness is not equivalent to the entity view or the idea that stem cells represent a natural kind. The scientific data highlighted in Chapter 6 lead to doubts both about whether stem cells represent a natural kind and about whether stemness is a categorical property. Although these data point to different issues underlying the conception of stem cells, they are oriented around two biological questions:

1. Cloning, iPS cells, STAP, regeneration, and cellular plasticity raise the following question: can cells dedifferentiate?
2. The niche hypothesis asks, is stemness controlled/determined by interactions between the cell and its niche?

The first question disputes the idea that stemness is an intrinsic property, because it asks if non-stem cells can acquire stemness. Therefore, it challenges the entity/natural kind view. The second question can either challenge the entity/natural kind view or adhere to this view and challenge the idea that stemness is a categorical property. I previously distinguished two interpretations of the stem cell niche hypothesis: the strong and the weak interpretations. The strong interpretation, according to which the niche determines stemness, refutes the conception of stemness as an intrinsic property and thus the entity/natural kind view because, according to it, the niche can induce stemness in non-stem cells. The weak interpretation, according to which the niche controls stemness, challenges the idea that stemness is a categorical property by disputing the idea that it exists in virtue of what stem cells are, without reference to the interactions that stem cells can have with their environment. This chapter focuses on this internal challenge to the entity/natural kind view, while the other questions are the subject of Chapter 8. Thus, the question I address in the following section is, what does the weak interpretation of the niche hypothesis tell us about stemness?

Stemness as a Dispositional Property

The weak interpretation of the stem cell niche maintains that interaction between stem cells and their niches determines stemness expression. The metaphysical tool that captures this type of property is the notion of "disposition"—wherein dispositions are properties of objects whose expression depends on extrinsic factors. This notion of disposition has two characteristics. First, as categorical properties, dispositions sort the objects into two categories: those that have the disposition and those that do not. For example, fragility is a dispositional property. Crystal glasses are fragile, but not iron cups. Second, in contrast to categorical properties, the expression of a disposition necessarily depends on extrinsic factors. A crystal glass will not break, unless it suffers a shock. The concept of disposition makes us consider the object-and-environment complex as the causal basis of the disposition. In other words, the environment plays an important causal role in the expression or nonexpression of the disposition.[1]

Applied to stem cells and CSCs, if stemness is a disposition, then stem cells belong to a natural kind whose expression of its essential property depends on interactions with the microenvironment. In other words, according to the supporters of the dispositional conception of stem cells, stemness is essential and specific to stem cells (as it is the case for proponents of the categorical view). Thus, stemness would rely on some intrinsic properties that only stem cells possess. But the expression of stemness depends on extrinsic factors, such as soluble factors that are located in the microenvironment, orientation of the mitotic spindle by the niche cell, or elasticity of the niche.

Some Data in Favor of the Disposition View for CSCs

The weak interpretation of the niche hypothesis presented in Chapter 6 has some support within the CSC research community. The existence and role of niches in cancers are debatable, and we still lack data, although it is an area of active research.[2] On one hand, cancers are often considered lineages of cells that no longer respond to environmental regulation; on the other hand, a number of studies suggest that there are significant specific interactions between cancer cells and their microenvironment (see, e.g., Schepers, Campbell, and Passegué 2015).

The relationship between cancer stem cells and the microenvironment is poorly understood, and the current lack of data makes any interpretation difficult. A study published by Christopher Calabrese and colleagues (2007) offers a good example of the ambivalence surrounding this metaphysical conception. Their research highlights that CSCs of brain tumors are dependent on a specific perivascular niche. The endothelial cells of the niche secrete factors that maintain the CSC in a self-renewal state *in vitro*. They observed that increasing the number of endothelial cells and/or blood vessels resulted in increasing the fraction of self-renewing cells and in the acceleration of the initiation and growth of tumors. Conversely, the depletion of blood vessels by anti-angiogenic treatments seemed to stop the growth of the tumor and eradicate the CSCs. Yet anti-angiogenics are at the heart of a controversy since Max Wicha's team published work showing that treatment of breast cancers with sunitinib or bevacizumab (two anti-angiogenics) increased the population of CSCs (Conley et al. 2012). This study is no less ambivalent than the work of Calabrese—it suggests both that CSCs are independent from the perivascular niche and that CSCs could respond to the hypoxic niche. They show that the hypoxic niche can enrich the CSC population thanks to hypoxia-inducible factor-1 alpha (HIF-1α) transcription factors.[3]

According to a review by Stéphanie Cabarcas, Lesley Mathews, and William Farrar, CSC activity could be regulated by:

- endothelial cells via vascular endothelial growth factor (VEGF);
- mesenchymal stem cells that produce cytokines such as stromal-derived growth factor-1 (SDF-1) and interleukin-6 and interleukin-8 (IL-6 and IL-8), which interact with CSCs' surface proteins;
- soluble factors present in the extracellular matrix of CSCs, such as ROS, HIF-1α, or transforming growth factor–beta (TGF-β); and
- lymphocytes via tumor necrosis factor–alpha (TNFα) or nuclear factor kappa beta (NF$\kappa\beta$), both inflammatory cytokines (Cabarcas, Mathews, and Farrar 2011; see also the review by Ye et al. 2014).

In 2008, Ilaria Malanchi and colleagues, at the Ecole Polytechnique Fédérale de Lausanne in Switzerland, highlighted the key role of β-catenin in the maintenance of CSCs in a skin cancer (squamous cell carcinoma). Four years later, they showed that the periostin secreted by stromal fibroblasts of the lung was required for circulating CSCs to metastasize. Their data suggest that without periostin, CSCs are unable to initiate secondary tumors by themselves (Malanchi et al. 2008, 2012; see also Wang and Ouyang 2012).

Together, these data suggest that CSCs, like stem cells, reside in niches that would control their division (self-renewal versus differentiation). Additionally, the niche seems able to determine the developmental fate of CSCs. James C. Mulloy's team, at the Cincinnati Children's Hospital Medical Center, showed that in leukemia, a clonal population of leukemic stem cells can develop into an acute myeloid leukemia or an acute lymphoid leukemia, depending on the microenvironment (Wei et al. 2008).

CSCs might also be able to create or modify their own niches. Research on cell competition between populations of cancer cells and populations of non-cancer cells has drawn attention to the ability of CSCs to act on their niches. According to Bin Zhang and colleagues, leukemic cells produce cytokines that modify the environment and subsequently confer a growth advantage to leukemic stem cells over hematopoietic stem cells (Zhang et al. 2012; see also Huang and Zhu 2012). In some brain cancers (gliomas), according to a review published by Alina Filatova, Till Acker, and Boyan K. Garvalov (2013), CSCs play a dominant role in promoting new blood vessel formation, thus generating new CSC-maintaining niches. Niche modification is a process that also seems to be involved in the production of metastatic niches. Expression of periostin, which Malanchi et al. (2012) showed to be necessary to maintain the CSCs, can be induced in the metastatic niches by infiltrating CSCs. CSCs may therefore not be passive in their relationship to their niche. The assumption that CSCs could build their niche, producing a microenvironment favorable to them, is an important argument in favor of the dispositional conception of stemness, insofar as it suggests the existence of some intrinsic causal bases that would be specific to CSCs.

So far, I have distinguished two possible stem cell identities: either stemness is a categorical property, or it is a dispositional property. This metaphysical distinction is important for biology, or at least for cancer research and the CSC theory, because different therapeutic strategies for cancers will lead to definitive cures depending on whether stemness is a categorical property or a dispositional property.

What if Stemness Is a Categorical Property?

Chapter 2 showed that the CSC theory derives from the existence and causal role of cancer stem cells and results in therapeutic strategies that specifically target CSCs. Recall that CSCs are defined by having stemness; that stemness is considered specific, among the cancer cells, to CSCs; and that this property specifically confers to CSCs the ability to develop cancers. As a consequence, it appears necessary and sufficient to kill all the CSCs to cure cancers. The other cells, insofar as they lack stemness, lack the tumorigenic abilities and are doomed to disappear.

Analyses developed throughout the four previous chapters raise several questions about the therapeutic strategy of CSC targeting. First, there is the question of whether this therapeutic strategy can overcome the heterogeneity of cells to which the concept of CSC refers (see Chapters 3–5). Chapter 5 showed that the CSC referent is open (see Figure 5.3) and can include the following:

- The stem cell initially mutated but still precancerous
- The mutated cancer stem cell initiating the original cancer and its daughter cancer stem cells that are prone to clonal evolution
- The cancer stem cells that are able to metastasize and to initiate a new cancer in a secondary organ
- The cancer stem cells that have survived the selection pressure imposed by therapy
- The cancer stem cells isolated in the laboratory that can initiate cancers in immunodeficient mice

Anti-CSC therapies must at least target the initiating CSC and its evolving daughter CSCs, the metastasizing CSCs, and the therapy-resistant CSCs (when appropriate) to warrant a cure. Precancerous CSCs might also be important since they represent a reservoir for new carcinogenesis (Corces-Zimmerman et al. 2014; Shlush et al. 2014; Ford et al. 2015). Thus, the question arises, are CSCs a single target or do CSCs constitute a range of targets?

Two kinds of solutions must be distinguished concerning this difficulty: theoretical solutions and empirical solutions. So far, the empirical solutions notably differ from the theoretical ones.

The theoretical solution is quite simple: if stemness is a categorical property, then it guarantees a more fundamental identity to the plurality of CSCs. Thus, targeting stemness would resolve the problem of heterogeneity among CSCs. Targeting stemness relies on the identification of some kind of molecular signature (either genetic or epigenetic) that constitutes this property. Such a signature has been shown to be controversial because neither self-renewal nor differentiation are specific to stem cells (Chapter 6). However, being too inclusive does not represent a major problem for CSC-targeting strategies as far as it allows the therapy to target every CSC of a given cancer. Targeting more than only the CSCs does not compromise the goal or the predicted cure. The major challenge for proponents of the CSC theory is distinguishing CSCs from normal stem cells (i.e., non-cancerous stem cells). Destruction of the patient's stem cells would seriously compromise their survival. Therefore, the therapy must eliminate the CSCs while preserving the non-cancerous stem cells. Thus, targeting CSCs through their stemness presents important challenges.

At least two theoretical solutions to this problem are possible. First, a specific target must distinguish CSCs from normal stem cells—this is the more classical strategy currently under development. Biologists try to identify markers that are overexpressed in CSCs compared to normal stem cells. Second, nanovectors can be designed to deliver the drug into the tumor. Vectorized drugs are drugs encapsulated in nanoparticles whose objectives are twofold: to preserve the active substance from interactions with the body until the target is reached and to avoid off-target effects on non-cancerous cells by delivering the drug at the site of the cancer. The nanovectors should deliver the treatment to the cancer cells and, in so doing, protect the non-cancerous stem cells. A few research teams have started this type of research (Vinogradov and Wei 2012;

Wang et al. 2013; Gul-Uludag et al. 2014; Kapse-Mistry et al. 2014). Given these solutions, if we assume that stemness is a categorical property, then with a given patient, if a drug succeeds in targeting and killing every CSC, then the CSC theory predicts a permanent cancer cure for that patient.

Empirically, things are quite different. Many researchers try to find targets, either surface marker proteins, signal cascades, or ABC cassettes, for CSC-targeting strategies.[4] Research generated by Irving Weissman's team provides a good example. Weissman's research is probably the best example of adoption of the categorical property conception of stemness, and he is a strong proponent of the anti-CSC therapeutic strategy. As of October 2014, his team was in the midst of conducting two clinical trials. One, called "Microarray Analysis of Gene Expression and Identification of Progenitor Cells in Lung Carcinoma," aims to identify the CSCs of carcinomas and secondary metastatic lung cancers through gene expression profiles.[5] The other one, titled "A Comprehensive Study to Isolate Tumor-Initiating Cells from Human Epithelial Malignancies," is described as follows:

> We hypothesize that all human malignancies harbour a subpopulation of tumor initiating cells/cancer stem cells (CSCs) that drives tumor development and potentially recurrence or metastasis of the disease. The primary aim of this study is to develop strategies for prospective isolation/enrichment of CSCs from human tumors of different tissue origins. In addition, we will characterize the signaling pathways and/or tumor specific antigens that are specific for CSCs, in order to specifically target these CSCs as the endpoint of this study.[6]

Weissman also founded the biotechnology company, Cellerant, which is developing "a Cancer Stem Cell (CSC) antibody discovery program focused on therapies for acute myeloid leukemia, multiple myeloma and myelodysplastic syndrome."[7] In these examples, we can see that the research programs fit the theory: to find a molecular signature to specifically target cancer stem cells. However, in practice, the higher challenge they face is not to distinguish CSCs from cancer non-stem cells but to distinguish CSCs from normal stem cells. There, the candidate molecular signature appears empirically founded, but with no demonstration that the studied molecular traits are present in *every* CSC. Currently, Weissman's team has pinned its hopes to the antigen CD47, on which they have worked for more than a decade. CD47 is a cell surface protein known for sending a "don't eat me" signal to the immune cells, therefore avoiding phagocytosis (Jaiswal et al. 2010). CD47 is not specific to CSCs, and its overexpression is a well-known phenomenon in cancers; however, hope for it as a CSC marker has risen since Weissman's team demonstrated that it was overexpressed in leukemic CSCs compared to hematopoietic stem cells (Majeti et al. 2009; see also Jaiswal et al. 2009). *In vitro* treatments of leukemia samples with CD47 monoclonal antibodies resulted in a considerable rise of phagocytized CSCs. The same treatment applied to hematopoietic cells did not raise the level of phagocytosis compared to control populations. Overexpression of CD47, then, appeared to be a potential molecular signature of leukemic CSCs.

The story of CD47 is interesting for two reasons. First, the overexpression of the CD47 antigen distinguishes the CSCs from the normal stem cells, but, among the cancer cells, it is far from being specific to CSCs. Overexpression of CD47 is a mechanism of defense against the immune system that numerous cancers develop—this is why Weissman's team developed anti-CD47 antibodies in the first place. Second, there is no experimental evidence that *all* CSCs overexpress the CD47 antigen. Weissman's group only showed that some CSCs are phagocytized after treatment (e.g., Willingham et al. 2012). Therefore, the hope that the development of CD47 monoclonal antibodies will fulfill the anti-CSC therapeutic strategy relies on the assumption that CSCs represent a homogeneous population, at least in regard to this character. But it remains a very weak assumption insofar as the relationship between CSCs and the CD47 antigen is unknown.

Research on CD47 is also a good example of how the CSC theory regenerated a lost hope for a universal therapy. Weissman's team found that CD47 is overexpressed in CSCs compared to stem cells in a variety of cancers: bladder cancers (Chan et al. 2009), non-Hodgkin's lymphoma (Chao et al. 2010), smooth muscle sarcoma (leiomyosarcoma) (Edris et al. 2012), multiple myeloma (Kim et al. 2012), or ovarian, breast, and colon cancers (Willingham et al. 2012). This success, currently limited

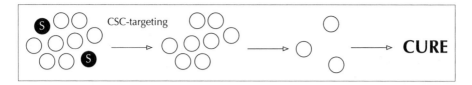

Figure 7.1. Evaluating the CSC-targeting strategy in light of the stem cell identity. If stemness is a categorical property, then it should be possible to target the CSCs and, in so doing, to cure cancers. Black S dots represent the cancer stem cells. White dots represent the cancer non-stem cells. (Redrawn from Laplane 2015, Figure 28.1. © 2015 River Publishers.)

to *in vitro* studies and *in vivo* transplantation into mice, allowed Weissman, Arash Alizadeh, Mark Chao, and Ravindra Majeti to file a patent for a synergistic anti-CD47 therapy for hematologic cancers (Weissman et al. 2012). The team also received a grant of $20 million from the California Institute for Regenerative Medicine to initiate a phase I clinical trial (Ricciardi 2012).

In summary, the CSC-targeting strategy of the CSC theory relies on the traditional categorical conception of stem cells. If stemness is a categorical property, then it should be possible to find a way to target the CSCs and, in so doing, to cure cancers (Figure 7.1). However, given the current absence of rigorous knowledge about stemness at the molecular level, practice remains relatively distant from theory.

What if Stemness Is a Disposition?

The "dispositional conception" of stemness does not challenge the CSC-targeting strategy. Indeed, if stemness is a disposition, as well as if it is a categorical property, the elimination of *every* CSC in a given cancer is necessary and sufficient to definitively cure the patient, since in both cases, only the CSCs are tumorigenic. The dispositional conception also aligns with the categorical conception on the existence of *some* intrinsic causal bases for stemness. In the dispositional view, stemness is not reducible to those causal bases. But only CSCs have those causal bases, and therefore, only CSCs can express stemness. Consequently, at least theoretically, the causal bases of stemness could be identified and targeted to eliminate CSCs and to cure cancers (Figure 7.1).

However, the dispositional conception of stem cells has two specific consequences for therapies. It suggests the existence of two new

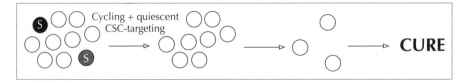

Figure 7.2. Evaluating the CSC-targeting strategy in light of the stem cell identity. If stemness is a disposition, then it should be possible to target the CSCs and, in so doing, to cure cancers. However, to cure the cancer, it is necessary to target all the cells that have the stemness disposition whether they express it (black S cell) or not (gray S cell). (Redrawn from Laplane 2015, Figure 28.1. © 2015 River Publishers.)

targets compared to the categorical conception: quiescent CSCs and the CSC niche.

First, the dispositional conception draws attention to the fact that the CSCs might express different molecular signatures over time, depending on their activity (quiescent or cycling). Peter Quesenberry, Gerald Colvin, and Jean-François Lambert (2002) developed an interesting artistic metaphor with the chiaroscuro painting technique to describe the molecular profile of stem cells. Chiaroscuro is a painting technique that involves using strong contrasts of light and shadow to hide and reveal an object. Both light and shadow are necessary to picture the object; without one or the other, we would only have a partial representation of the object. They used the chiaroscuro metaphor with respect to stem cells to indicate that a molecular portrait of stem cells based on cell division (self-renewal and differentiation, both implying cell division) would be tantamount to removing the dark from a chiaroscuro painting.

> In essence, the identity of the stem cell would be masked at certain points in the cycle, and in a nonsynchronized population of cells, at any one point in time, a number of true stem cells would not be detectable. (Quesenberry, Colvin, and Lambert 2002, 4270)[8]

This metaphor points out that it might be more efficient to target separately the quiescent and the cycling CSCs (see Figure 7.2). Remember the example of the CD47 target identified by Weissman's team. This target is poorly specific to CSCs, which means that it is not specific to the categorical bases of stemness. Within the dispositional view, because CD47 is not related to the categorical bases of stemness, it might not be expressed by every CSC. And indeed, some data suggest that it is the case with hematopoietic stem cells (HSCs). HSCs express CD47 when the hematopoietic system actively recruits them, but not (or poorly) when they are in their niche, where they are often quiescent. Consequently, if CSCs can also be quiescent, as some studies suggest (see Moore and Lyle 2011), there is a risk that quiescent CSCs do not express CD47. If this is true, then these CSCs will escape anti-CD47 therapies, and the CSC theory predicts a high risk of relapse.

As a disposition, stemness might rely on some intrinsic characters (to which it cannot be reduced). Thus, at least in theory, it is possible that a single complex signature distinguishes between CSCs and cancer non-stem cells. For example, the review by Cabarcas et al. (2011) mentioned earlier suggests that cell surface receptors such as VEGFR, CXCR4, IL-6, and IL-8 make CSCs receptive to the stimuli sent by the niche. These receptors could be good candidates for a molecular signature of the causal bases of the stemness disposition and therefore potential targets for CSC-targeting therapies. Remember the example of fragility: with the same stimuli (a fall), the crystal glass breaks whereas the iron cup does not. This is because there is something (some causal bases) in the crystal glass that makes it receptive to this environmental stimulus. By analogy, it is possible to hypothesize that receptors that make CSCs, but not other cancer cells, receptive to niches' stimuli might be related to the causal bases of stemness. In any case, as mentioned with the categorical conception, the most important point is to verify the ubiquity of the chosen signature(s) to guarantee a definitive cure.

The dispositional conception raises another possibility to cure cancers. Metaphysicians have debated for a long time on how to frame what a disposition is. The classical solution, called the "simple counterfactual analysis," follows:

> Something x is disposed at time t to give response r to stimulus s if and only if, if x were to undergo stimulus s at time t, x would give response r. (Lewis 1997, 143; Bird 1998, 227)

For example, fragility of the crystal glass can be said to be a disposition of the crystal glass. That is to say that the crystal glass (x) is disposed at time *(t)* to break (response *r*) in case of some impacts (stimulus *s*), because if the crystal glass *(x)* were to undergo an impact at time *t*, it would break (response *r*). Philosophers have suggested diverse counter-examples to discuss this formulation of dispositions. Among these counterexamples, one is particularly interesting for us—the "antidote" of Alexander Bird (Bird 1998; see also Choi 2003). Applied to the crystal glass example, this counterexample takes the following form: the crystal glass is called fragile because it would break if impacted. But if a box filled with foam was protecting the crystal glass, then the simple conditional counterfactual analysis would become false. The protective equipment applied to the crystal glass breaks the causal chain leading from the stimulus *s* (impact) to the response *r* (crystal glass breaking).

This counterexample offers a pertinent analogical model for the development of anti-cancer therapies. Indeed, any therapy that would break the causal chain leading from the niche stimuli to the expression of stemness would represent a good "antidote." According to the CSC theory, CSCs are tumorigenic (i.e., can initiate, maintain, and develop cancers) because of their stemness disposition. Without the expression of stemness, there would be no cancers or metastases. The dispositional view highlights that getting ride of the disposition is not the only option. The other option is to prevent its actualization. The weak interpretation of the niche hypothesis claims that the expression of stemness (response *r*) follows from stem cell niche stimuli *(s)*. Thus, targeting the stem cell-niche relationship would make the CSCs non-tumorigenic. Such a strategy should have the same outcome as the CSC-targeting therapies, that is, a definitive cure for cancers (see Figure 7.3). The main challenge to develop a therapy that would be analogous to an antidote in the relationship between *r* and *s* is the lack of data regarding the stimuli produced by the niche that is necessary for CSCs to express their stemness disposition. This kind of treatment strategy is therefore often very theoretical in publications (e.g., Clarke and Fuller 2006; Sneddon and Werb 2007; Federici et al. 2011; Huang and Zhu 2012).

Despite the difficulties, the scientific literature offers some interesting avenues of research along these lines. For example, Marieke Essers and Andreas Trumpp (2010) have suggested a two-step protocol to eliminate quiescent and non-quiescent CSCs. The first step is to activate quiescent CSCs, and the second step is to perform an anti-CSC

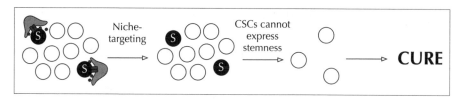

Figure 7.3. Evaluating the CSC-targeting strategy in light of the stem cell identity. If stemness is a disposition, then either we can target the disposition (as in Figure 7.2) or we can target the extrinsic stimuli that control the expression of the disposition—namely, the CSC niche (gray cells with semicircle shape). If CSCs cannot express their stemness, then they will not be able to maintain the cancer, which will regress, and lead to a cure. (Redrawn from Laplane 2015, Figure 28.1. © 2015 River Publishers.)

targeted therapy. To activate quiescent CSCs, they consider several solutions consisting of producing stimuli *s* that would mimic stimuli of the niche to obtain the desired result *r*—namely, the activation of the quiescent CSCs. For example, in bone marrow, the release by the immune system cells, called neutrophils, of proteolytic enzymes, such as matrix metalloproteinases that cleave and degrade the adhesion proteins that retain stem cells in their niche, can result in the mobilization of hematopoietic stem cells out of their niche (Heissig et al. 2002; Petit et al. 2002). Equivalent treatment for cancers could activate the quiescent CSCs. Several studies indicate that a treatment with granulocyte–colony-stimulating factor (G-CSF) could help achieve such results for chronic myeloid leukemia (CML) (Drummond et al. 2009) and acute myeloid leukemia (AML) (Saito et al. 2010). Pier Paolo Pandolfi's team, at the Memorial Sloan-Kettering Cancer Center, in New York, suggested a treatment with arsenic trioxide (As_2O_3) that targets the transcription factor promyelocytic leukemia protein (PML), whose inhibition leads to the activation of leukemic CSCs (Ito et al. 2008; Ito, Bernardi, and Pandolfi 2009). Essers and Trumpp also suggested a treatment based on interferon alpha (INFα). INFα was used for the treatment of CML before it was replaced by imatinib in 2001. Along with their colleagues, Essers and Trumpp showed that the application of INFα activated the division of hematopoietic stem cells (Essers et al. 2009). The effect of INFα on CSCs remains unknown. However, the results of the clinical trial for imatinib in France raise hope for the use of such drugs: over two years, six patients were in complete remission from CML, and all six were previously treated with INFα before they entered the clinical study (Rousselot et al. 2007).

Malanchi and colleagues (2008) offer another example. They showed that β-catenin binding protein is required to maintain the CSCs in squamous cell carcinomas. Breaking the niche activation of this pathway should allow the depletion of the stock of CSCs and thus lead to the regression and disappearance of the cancer.

As noted by Christopher Calabrese and his colleagues, these works all face the same difficulties—they risk the destruction of non-cancer stem cells: "The development of anti-CSC therapies for each type of cancer is likely to require the identification of factors that maintain CSCs, but not normal stem cells, in each tissue" (Calabrese et al. 2007, 69). This problem is, however, considered in most work. Malanchi et al. (2008) showed, on one hand, that β-catenin is preferentially expressed by CD34$^+$ cancer cells, among which there are the CSCs, and on another hand, that the elimination of CD34$^+$ cells does not affect the homeostasis of the healthy epidermis. Similarly, Essers et al. (2009) indicate that the therapeutic molecules that they use preferentially affect CSCs.

In addition, several molecules seem able to specifically target CSCs compared to non-cancer stem cells. The immunosuppressive agent rapamycin, which inhibits the mTOR protein, has been considered a proof of concept (Sneddon and Werb 2007). Sean Morrison's team showed that the tumor suppressor gene *Pten,* whose deletion leads to an increased activation of *Akt* and *mTOR,* regulates hematopoietic stem cells and leukemic CSCs in different ways (Yilmaz et al. 2006). mTOR would therefore represent a CSC-specific therapeutic target in leukemia. Using a Cre-Lox mouse model—i.e., one in which a gene (here *Pten*) can be deleted at a given time and in a given tissue—Morrison's team showed that the deletion of *Pten* leads to the depletion of hematopoietic stem cells and promotes the generation of leukemic CSCs. They also showed that treating the mice with rapamycin inhibits the generation and maintenance of leukemic CSCs and restores the hematopoietic stem cells. Although rapamycin treatment is not sufficient to cure leukemia, the authors wrote, "These data demonstrate that it is possible to identify—and to target therapeutically—pathways that have distinct effects on normal stem cells and cancer stem cells within the same tissue" (Yilmaz et al. 2006, 481).

CONCLUSION

In this chapter, tools borrowed from metaphysics allowed me to establish a distinction between two possible stem cell identities. Stemness can be either a categorical or a dispositional property, each of which has repercussions for cancer therapies.

The traditional conception of stem cells, on which many current CSC-targeting therapeutic strategies rely, considers stemness to be a categorical property. That is, stemness is a property that is essential and intrinsic to the cells belonging to the stem cell natural kind. Applied to the CSCs, if stemness is a categorical property, then it must be possible to target it and, thus, to eliminate the CSCs. Furthermore, lack of specificity of the biological markers for stemness is not a problem as long as they allow elimination of all the CSCs. The major empirical difficulty for the development of CSC-targeting drugs with stemness as a categorical property is distinguishing between CSCs and normal stem cells to kill only the former.

The dispositional conception gathers together the proponents of the weak version of the niche hypothesis presented in Chapter 6. According to this conception, the stemness property relies on some categorical bases that are specific to stem cells, but its expression depends on the intervention of extrinsic factors. That is, stemness also depends on the niche. If stemness is, indeed, a dispositional property, then:

1. CSCs might be more easily targeted through two distinct molecular signatures (one for the quiescent CSCs, the other for the cycling CSCs).
2. CSC-niche interactions represent another interesting target for therapies. By disrupting this relationship, a therapy would make the CSCs non-tumorigenic. This should have the same outcome as directly targeting the CSCs, i.e., a definitive cure for cancer.

8

IF STEMNESS IS A RELATIONAL
OR A SYSTEMIC PROPERTY,
HOW CAN WE CURE CANCERS?

In Chapter 7, I established a distinction between two possible identities for stem cells: stemness can be either a dispositional property (niche dependent) or a categorical property (niche independent). In both cases, stemness is considered an intrinsic property of the cells. If stemness is an intrinsic property (either dispositional or categorical), then the therapeutic model of the CSC theory described in Part I applies: eliminating CSCs should be necessary and sufficient to cure cancers. In this chapter, I show that this CSC-targeting therapeutic strategy will not be effective if stemness is an extrinsic property that non-CSCs can acquire (i.e., the "state," "function," or "nurture" view). Within the category of stemness as an extrinsic property, two more possible identities for stem cells exist: stemness can be either a relational property (niche dependent) or a systemic property (niche independent). The consequences of both possibilities for cancer treatments are discussed.

IF STEMNESS IS AN EXTRINSIC PROPERTY, TARGETING CSCs MIGHT FAIL TO CURE CANCER

Chapter 6 reviewed data suggesting that cells can dedifferentiate and acquire stemness. If this is true, then stemness is not an intrinsic property specific to a defined population of cells but an extrinsic property that non-stem cells can acquire. The possibility that stemness might be an extrinsic property has dramatic consequences for the model of targeted therapy proposed by the CSC theory. To understand the consequences of these different understandings of stem cell identity, I suggest a metaphor from the *Matrix* movies by the Wachowski brothers.

In the *Matrix* trilogy, machines "cultivate" humans in towers, providing them liquid nutrients and oxygen, in order to feed themselves with bioelectricity. To keep the humans alive and producing energy, a complex computer program called the Matrix offers them a form of virtual life on Earth that gives them the illusion of freedom. Within the Matrix, agents ensure stability of the program and fight against humans who try to hack it. At any time, the agents can integrate into any human body wired to the Matrix. When they do so, the human is transformed into an agent, absorbing its function and properties. For example, agents have an increased mobility and speed of movement.

Being an agent is a state, or an identity, defined by extrinsic properties. Agents represent a possible reversible transient state for any human of the Matrix. In this metaphor, "agent" is to human what "stemness" is to (cancer) stem cells if stemness is an extrinsic property. Stretching the metaphor, what would happen in the Matrix if one would develop the equivalent of the CSC-targeting therapy, that is, if someone killed all the agents of the Matrix? In the movies, when an agent is killed, the human who was the agent dies, but the "agent" property survives. Another human acquires the property and becomes the agent. Thus, if stemness is a property of that kind, killing all the CSCs of a given cancer at time *t* cannot ensure the recovery of the patient. At any time, a cancer non-stem cell can acquire stemness and generate a relapse (see Figure 8.1).

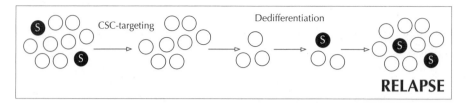

Figure 8.1. Evaluating the CSC-targeting strategy in light of the stem cell identity. If stemness is an extrinsic property, then targeting CSCs (black S cells) might not be sufficient to cure cancers because cancer non-stem cells (white cells) can acquire stemness and cause a relapse. (Redrawn from Laplane 2015, Figure 28.1. © 2015 River Publishers.)

The question of stem cell identity is far more than a philosophical question attracting the interest of biologists who enjoy conceptual and theoretical debates. It is crucial to determine what kind of property stemness is to adopt the right therapeutic strategy and to decide where to invest funding and research efforts. Thus, the question of stem cell identity requires discussion and movement toward resolution for both the future of patients suffering from cancer and the researchers and funders, including pharmaceutical industries that fund the clinical trials and market the drugs.

Two questions arise from the observation discussed in Chapter 6 that non-stem cells can dedifferentiate to the stem state. First, is dedifferentiation a plausible phenomenon in cancers? In other words, is dedifferentiation an exceptional phenomenon or does it represent a real risk in cancer? Second, if dedifferentiation is a real risk, what are the therapeutic options to fight against cancers?

CAN CANCER NON-STEM CELLS DEDIFFERENTIATE TO BECOME CSCs?

The idea that cancer non-stem cells can acquire stemness arose in 2006. Janusz Rak, at the Montreal Children's Hospital Research Institute, Canada, formulated the hypothesis that stemness would be a transient and acquirable property.

> It should be considered (and it has been in the literature), that the "stemness" of the putative subset of cancer initiating cells may not be permanent, but rather could possibly

be a transiently acquired property, not necessarily absolute (qualitative), but instead relative (quantitative) in nature. (Rak 2006, 602)

In addition, based on the observation that numerous factors can influence the ability of transplanted cancer cells to graft and develop cancers, Rak suggested that stemness was not attributed to the right unit. According to him, stemness should be attributed to "interactive *multicellular cancer stem units*"—i.e., collections of cells, instead of individual cells. "The inner circuitry of such units (not of one cell type, in particular) would, therefore, be responsible for tumor take, repopulation and maintenance" (Rak 2006, 603). This model recalls Ray Schofield's stem cell niche hypothesis. In both cases, stemness is assigned to a cell-plus-environment complex instead of to cells at the individual level, and the loss of stemness is reversible. This is what I called the strong interpretation of the niche hypothesis.

Concrete data in favor of the idea that cancer cells can dedifferentiate and acquire stemness also emerged from several groups working on different cancers.

Robert Weinberg's team at MIT identified a subpopulation of mammary epithelial basal cells in humans capable of spontaneous dedifferentiation. They demonstrated that the probability of such a dedifferentiation increases in breast cancers (Chaffer et al. 2011).[1] Furthermore, Weinberg's team showed that epithelial to mesenchymal transition (EMT), a process during which epithelial cells become mesenchymal cells and lose their cell polarity and cell-cell adhesion (thus gaining migratory and invasive properties), could induce the emergence of stemness characteristics in both normal and cancerous breast tissues (Mani et al. 2008). The cells (either cancerous or not) that underwent an EMT express the molecular markers of undifferentiated cells. They show an increased capacity to produce mammospheres, the functional capacity by which biologists usually identify mammary stem cells because it highlights the capacity of the cell to self-renew and differentiate. Finally, Weinberg's team showed through both *in vitro* and *in vivo* experiments that these breast cells that undergo EMT have an ability to produce tumors (tumorigenicity) 10 times higher than the cancer non-stem cells if the signals inducing the EMT are maintained (see also

Scheel et al. 2011). Work of a team led by Alain Puisieux in Lyon corroborated Weinberg's findings. Puisieux's group generated CSCs from mammary epithelial cells by activating the Ras-MAPK signaling pathway, making cells more sensitive to undergoing the EMT (Morel et al. 2008). More recently, Weinberg's group observed that the probability that a cancer non-stem cell will dedifferentiate into a CSC varies among different breast cancer cell lines. Basal-type breast cancer cell lines show a great aptitude for dedifferentiation, whereas almost no dedifferentiation happens under the same conditions with the luminal types (Chaffer et al. 2013).

Jan Paul Medema's team at LEXOR, in Amsterdam, studied the role of Wnt signaling pathways in colorectal CSCs. They showed that some factors produced by the CSCs' niche cells—the hepatocyte growth factor (HGF) produced by myofibroblasts—regulate the Wnt pathway. Their results suggest that HGF could induce the dedifferentiation of cancer non-stem cells into CSCs.

> We show that differentiated cancer cells, which have lost the capacity to form tumours and are no longer clonogenic, can be reprogrammed to express CSC markers and regain their tumorigenic capacity when stimulated with myofibroblast-derived factors. This suggests that cancer stemness is not a rigid feature but can be modulated and even installed by the microenvironment. (Vermeulen et al. 2010, 468)

Following from this observation, they suggested replacing the hierarchical model of CSCs with what they call the "dynamic CSC model," represented in Figure 8.2 (Vermeulen et al. 2012, e84). According to this model, the niche controls the CSCs' activities and can also induce dedifferentiation of cancer non-stem cells.

Moreover, a team led by Eric Lander, one of Weinberg's colleagues at MIT, studied the dynamics of cancer cell subpopulations in a breast cancer cell line (Gupta et al. 2011). Their results suggest that dedifferentiation may be a stochastic phenomenon. After using FACS to sort cells into three subpopulations, one enriched in CSCs, one enriched in basal cells, and one enriched in luminal cells, they cultivated the subpopulations *in vitro*. Following cultivation, each subpopulation returned to the equilibrium—that is, the cultures developed the same ratios of basal

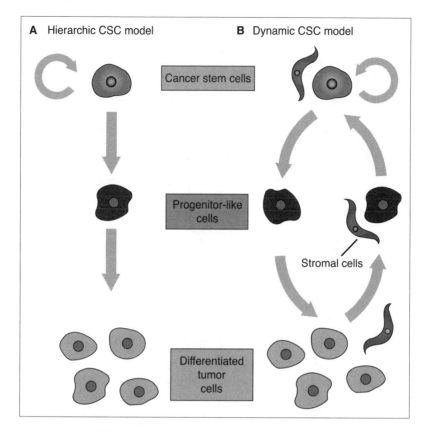

Figure 8.2. The hierarchical CSC model and the dynamic CSC model represented by Vermeulen et al. (2012, e84). (Reprinted with permission from Vermeulen et al. 2012, Figure 1. © 2012 Elsevier Ltd. Reprinted with permission from Elsevier.)

cells, luminal cells, and stem cells that had been in the cell population pre-FACS. Lander's group used Markov's model of stochastic transitions between states to explain this phenomenon. This model allows two predictions, both corroborated by the results of their experiments:

> A prediction of this model is that, given certain conditions, any subpopulation of cells will return to equilibrium phenotypic proportions over time. A second prediction is that breast cancer stem-like cells arise de novo from non-stem-like cells. (Gupta et al. 2011, 633)

On the basis of these results, Lander proposed an alternative model of cancer development, where "bidirectional interconversions" are possible

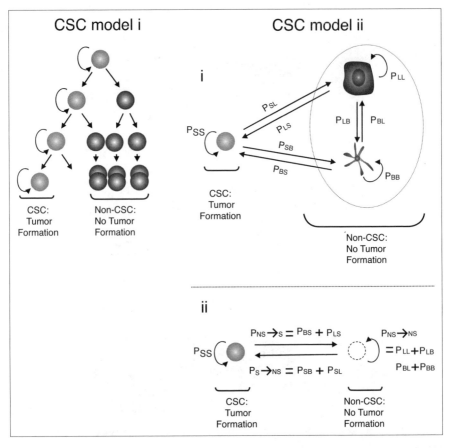

Figure 8.3. The hierarchical CSC model and the bidirectional interconversion CSC model represented by Gupta et al. (2011, 641). The rates of transitions between two cell states can be computed with the Markov modeling approach. (Reprinted with permission from Gupta et al. 2011, Figure 5. © 2011 Elsevier Inc. Reprinted with permission from Elsevier.)

and predictable between the stem state and other non-stem states (see Figure 8.3, from Gupta et al. 2011, 641). In their review titled "Cancer Stem Cells: Impact, Heterogeneity, and Uncertainty," Jeffrey Magee, Elena Piskounova, and Sean Morrison (2012) suggested a similar bidirectional interconversion CSC model, and another mathematical model was framed by Da Zhou et al. (2013).

In accordance with the studies by Weinberg's, Medema's, and Lander's team on colon and breast cancers, Yugang Wang's team at the State Key Laboratory of Nuclear Physics and Technology in Beijing observed that non-CSCs of some colon and breast cancer cell lines could spontaneously express a cell surface marker (CD133) usually associated with CSCs, suggesting that non-CSCs can transform into CSCs (Yang et al. 2012).

Thomas Tüting's group, at the University of Bonn, has reported that dedifferentiation plays a role in melanoma resistance to adoptive cell transfer therapies. These therapies involve transferring immune cells into patients (in this case, T cells) to increase immune response against the cancer. Using such therapeutic strategies, patients with metastatic melanoma can enter remission; however, relapses are frequent (Rosenberg et al. 2011; Restifo, Dudley, and Rosenberg 2012). According to Tüting, dedifferentiation in these cases could be induced by the inflammatory microenvironment produced by the T cells, particularly through the pro-inflammatory factors such as TNFα (Landsberg et al. 2012).

Ulf Rapp, Fatih Ceteci, and Ralf Schreck (2008) have suggested a model of dedifferentiation wherein CSCs might be the passengers rather than the drivers of carcinogenesis. They argue that oncogenes induce plasticity, which results in the acquisition of stemness by some cancer cells. The major role of some oncogene such as *c-Myc* in the production of iPS cells is an important argument in favor of this model.

> Therefore we propose an alternative view: especially in tissues with slow turnover rates tumor initiating mutations may occur in functionally mature progenitor or even terminally differentiated cells. These cells subsequently become unstable with respect to their phenotypes and acquire features of progenitor or stem cells corresponding to organ-specific or even embryonic stem cells. Dedifferentiation of tumor initiating cells by oncogenes presumably also occurs in tumors arising in tissues with rapid turnover. This is for example the case when the first mutations hit progenitor cells and may constitute the critical step in progression to generalized disease and metastasis, if it extends beyond the stage of organ specific stem cells. (Rapp, Ceteci, and Schreck 2008, 45)

Research conducted by Marie-Pierre Junier's and Hervé Chneiweiss's group in Paris supports a possible dedifferentiation through cancer development in pediatric brain tumors. While their earlier work showed that TGFα growth factor, which can be highly expressed in gliomas, can induce dedifferentiation in astrocytes (Dufour et al. 2009), their more recent work indicates that some brain tumors seem able to develop without CSCs and that the progression of these tumors appears to induce the creation of CSCs (Thirant et al. 2011). Other studies have supported the idea that dedifferentiation occurs in glioblastoma (reviewed in Safa et al. 2015). Notably, Brenda Auffinger et al. (2014) performed lineage-tracing analyses that suggested that the expansion of glioblastoma CSCs following treatment with temozolomide, the most commonly used antiglioma chemotherapy, is a result of a conversion of non-CSCs into CSCs.

Cases of dedifferentiation of non-CSCs to CSC have also been reported in osteosarcomas (Zhang et al. 2013), pancreatic cancers (Zhu et al. 2013), and pancreatic ductal adenocarcinomas (Singh et al. 2015). Together, these data provide a sound basis for us to consider that dedifferentiation is a phenomenon in (at least some) cancers. Interestingly, the idea that cell dedifferentiation is common in cancers is not new. In the 1970s, the undifferentiated aspect of cancer cells was commonly explained by dedifferentiation (see Chapter 3). In cancers, dedifferentiation might be much more than an exception to the rule—a factor that could seriously limit the efficiency and efficacy of CSC-targeting therapies. If these results are confirmed and generalized, then which therapeutic options would be left under the CSC theory framework?

THERAPEUTIC STRATEGIES: WHAT ARE THE OPTIONS?

If stemness is an extrinsic property in cancers, such that cancer cells can dedifferentiate to produce new CSCs, then anti-CSC drugs could be used as supplementary therapies. In this case, the primary aim of cancer therapies would return to eliminating the highest number of cancer cells possible, and anti-CSCs drugs would be used in conjunction with other therapies to target cells (CSCs) that evade traditional therapies. This solution, however, faces a number of difficulties. First, any cancer cell that escapes the therapy could lead to a relapse if it acquires stemness. Combining a classical therapy with an anti-CSC therapy does not guarantee the elimination of *every* cancer cell. Second, the cost of such therapies could be extremely high in terms of monetary output as well as personal health.

The cost of therapies remains an underconsidered problem and deserves more attention. For instance, in a recent article on the cost of smart drugs, Tito Fojo and Christine Grady (2009) report that the anti-angiogenic bevacizumab (Avastin) costs around $8,000 per month per patient. For a patient with colorectal cancer, with an average survival of 18 months, the cost of a bevacizumab treatment is around $144,000, for an average increase of survival of four months. Personal health cost of the treatment for the patient can also be substantial—side effects of bevacizumab treatment include risk of gastrointestinal perforation, severe bleeding, severe arterial hypertension, blood clots, and delayed wound healing.

An alternative therapeutic option, based on the hypothesis of the stem cell niche, could be the disruption of the relationship between the niche and the stem cell. Proponents of both the weak and the strong interpretation of the niche hypothesis agree that the niche determines the ability of CSCs to initiate, develop, and propagate cancers. Accordingly, eliminating the niche stimuli could induce the differentiation of CSCs and could also avoid dedifferentiation of cancer non-stem cells.

Jan Paul Medema and Louis Vermeulen (2011) suggest several areas of possible therapeutic interventions for colon cancers based on niche disruption. In particular, their findings suggest that inhibition of c-Met receptors could prevent dedifferentiation. This is an adjuvant therapeutic strategy that comes in addition to either a CSC-targeting strategy (they suggest targeting CSCs through the bone morphogenetic protein [BMP] receptor) or to a two-stage strategy that includes differentiation (by blocking self-renewal through inhibitors of Wnt or Notch pathways, for example) followed by a classic chemotherapy. Patients undergoing either of these options face significant risk of side effects because the BMP receptor is not specific to CSCs, and Wnt and Notch pathways play important roles in normal stem cell self-renewal. If this type of strategy becomes prevalent, then nanovectors would represent a major area of research for the development of targeted therapies against cancer.

Disrupting the stem cell-niche relationship represents a plausible cure only if dedifferentiation is dependent on the niche (Plaks, Kong, and Werb 2015). Some of the data mentioned above indicate that dedifferentiation could occur independent of any specific stem cell-niche interaction. For example, the stochastic model of Gupta et al. (2011)

suggests that transitions between cell states can happen independent of the action of any specific niche. Robert Weinberg's team also considers this possibility: "We have shown that differentiated mammary epithelial cells can convert to a stem-like state, doing so in an apparent stochastic manner in vitro" (Chaffer et al. 2011, 7954). According to them, dedifferentiation can occur either spontaneously or through EMTs, a case of microenvironment-induced dedifferentiation.

The idea that dedifferentiation could appear through other mechanisms than niche-induced dedifferentiation is also present in scientific literature on normal stem cells, for example, in Arthur Lander's article on the stem cell concept:

> Stemness is a property of systems, rather than cells, with the relevant system being, at minimum, a cell lineage, and more likely a lineage plus an environment. A system with stemness is typically one that can achieve a controlled size, maintain itself homeostatically, and regenerate when necessary. (Lander 2009, 70)

According to these views, a system (or a cell line and its microenvironment) would be able to regulate equilibrium and homeostasis of the cell population through diverse means, among which niche signaling would be only one possibility.

The involvement of stochastic phenomena in determining the emergence of stemness could jeopardize the effectiveness of niche-targeting therapeutic strategies. If cancer non-stem cells are capable of acquiring stemness independent of any niche, then neither the CSC-targeting therapies nor the niche-targeting therapies can guarantee the recovery of patients suffering from cancer.

Therefore, it is important to distinguish two possibilities within the conception of stemness as an extrinsic property (the "state," "function," or "nurture" view): either stemness is specifically induced by a relationship with a niche, or it is induced by other factors. In the first case, stemness would be a "niche-induced property," that is, stemness would be what a metaphysician would call an "extrinsic relational property." In the second case, stemness would be a "systemic property" maintained, at the level of a given system, by a mixture of extrinsic and intrinsic stochastic factors.

If the niche is necessary for acquiring stemness, then the niche-targeting therapeutic strategy is a plausible alternative. But what kind of property is stemness if it is a "niche-induced property"? In metaphysical terms, it can be defined as an "extrinsic relational property." Because the distinction between intrinsic and extrinsic properties relates to the identity of the object with respect to itself or with respect to others, extrinsic properties and relational properties are often considered equivalent and opposite to intrinsic properties. For example, in 1983, David Lewis considered the concepts of relational and extrinsic properties as equivalent and defined them as follows:

> A thing has its intrinsic properties in virtue of the way that thing itself, and nothing else, is. Not so for extrinsic properties, though a thing may well have these in virtue of the way some larger whole is. The intrinsic properties of something depend only on that thing; whereas the extrinsic properties of something may depend wholly or partly, on something else. (Lewis 1983, 197; see also Lewis 1986, 61; Khamara 2006, 58)

Nonetheless, Lloyd Humberstone (1996) argued that the equivalence between relational and extrinsic properties is misleading because some intrinsic properties are relational, too. Stemness is a case of a property that could be either relational intrinsic or relational extrinsic. Indeed, according to the weak interpretation of the niche hypothesis, stemness is also a relational property in the sense that it depends on a specific relationship with the niche. In other words, both the weak and the strong interpretations of the niche hypothesis have the consequence that stemness is a relational property; in the former case, the property is intrinsic (see Chapter 7), whereas in the latter case, it is extrinsic.

In this regard, David Lewis's distinction between intrinsic and extrinsic properties might be useful: "If something has an intrinsic property, then so does any perfect duplicate of that thing; whereas duplicates situated in different surroundings will differ in their extrinsic properties" (Lewis 1983, 197). If one agrees with this distinction, then stemness is an intrinsic property in the weak version of the niche hypothesis

Table 8.1. Dispositional property versus relational-extrinsic property in light of duplicates.

	Weak version of the niche hypothesis (stemness is a dispositional property)	Strong version of the niche hypothesis (stemness is a relational extrinsic property)	
Object	*a* is a stem cell	*b* is a stem cell in a given niche *(n)* at a given time *(t)*	*c* is a non-stem cell
Duplicate	*a'* is a stem cell	*b'* is a stem cell if and only if it is located in niche *(n')*	*c'* is a stem cell if localized in *n'*

and an extrinsic property in the strong version of this hypothesis. Indeed, in the weak version, any duplicate *a'* of a stem cell *a* has the disposition to express stemness if it would meet the appropriate stimuli. But, whether or not they meet such stimuli, *a* and *a'* are considered stem cells (see Table 8.1, second column). However, in the strong version of the niche hypothesis, duplicate *b'* of a stem cell *b*, localized in a niche *n* and expressing stemness at time *t*, will not express stemness if it is not localized in a niche *n'* (Table 8.1, third column). Conversely, duplicate *c'* of a non-stem cell *c* could express stemness if it were located in an appropriate niche (Table 8.1, fourth column). Thus, in the weak version of the niche hypothesis (stemness as a dispositional property), if *a* is a stem cell, then *a'* is also a stem cell, whereas in the strong version, *b* can be a stem cell and *b'* a non-stem cell (and conversely). Furthermore, *b* can stop being a stem cell, and *c* can become a stem cell.

Gender in Melanesia offers a good example of an extrinsic relational property. According to anthropologist Marilyn Strathern's research, in Melanesia, gender is a property attributed by gift relations (Strathern 1988). A gift attributes "femininity" and "masculinity" properties to the giver and the recipient, respectively. In this culture, gender is not considered an attribute of the person. Gender does not depend on any intrinsic property but only on extrinsic relationships. It is a mobile and reversible property; one can be sometimes male and sometimes female depending on the gift-relation they have with others.

Similarly, characterizing stemness as an extrinsic relational property suggests that a particular niche-cell relationship determines whether a cell is a stem cell or not.

This characterization of stemness as an extrinsic relational property is not entirely satisfactory. It has the advantage of making intelligible the fact that the rupture of the cell-niche relationship determines the loss of stemness and provides a possible target for therapeutic strategies against cancers. But it has the inconvenience of not accounting for the functional aspect of the stemness property. Stemness, in contrast with other relational properties such as "being fifty miles south of a burning barn," involves specific actions.[2] Thus, an adequate metaphysical description of the niche-induced conception of stemness should both account for the relational aspects of stemness and for its functional aspects, that is, for the fact that possessing stemness allows a cell to act in certain ways (i.e., to self-renew and to differentiate).

STEMNESS AS A SYSTEMIC PROPERTY

If stemness can be acquired independently from any niche, as suggested by Gupta et al.'s (2011) experiments, then what kind of property is stemness? According to the data presented in this chapter, apart from the niche, at least two kinds of processes can induce stemness: stochastic events affecting gene expression and cell-population-level regulations. In both cases, stemness appears to be regulated at the population/system level, suggesting that stemness would be a "systemic property."

Stem cell system biology consists of considering cell identities as part of a cell state landscape, wherein identity is conferred by the cell's current position within that landscape (see Huang 2010, 2011, 2013; Ho et al. 2012). Stem cell system biologists use the epigenetic landscape metaphor developed by Conrad Waddington, in which a landscape of hills and valleys represents the states through which a cell passes as it differentiates (see the analysis of Fagan 2012). The peaks are the least differentiated states, and the cell gradually moves from them into a valley, which represents the most differentiated states. In such landscapes, some cell states are more stable than others: the bottoms of valleys are the most stable (called "attractors"), whereas the hills are highly unstable.

> The attractor state is a stable steady state and its "position"
> in state space is a characteristic of a system: it is the "set
> point" to which the system returns after perturbation. . . .
> This behavior is a property of a system and is encoded in its
> internal wiring. (Huang 2013, 428)

Usually, cell types are considered attractors, whereas stem cells are considered in a metastable state, a state in between two attractors (Huang 2011). However, stem cell system biologist Sui Huang, at the Institute for Systems Biology in Seattle, Washington, suggests that in the case of cancers, the stem state represents an attractor (Huang 2013; Pisco and Huang 2015).

The systemic conception of stemness differs from the three previous ones in that it accounts for stemness at a system level, rather than at the single cell (or niche-cell) level. The systemic conception of stemness is therefore characterized by a change of scale: "systems are taken to constitute a fundamental ontological category," from which it is possible "to understand the emergent properties of component interactions" (O'Malley and Dupré 2005, 1271). The interactions experienced by a cell determine its position in the landscape, that is, its state at time t. Saying that stemness is a systemic property means that the stem state is governed by internal and external stimuli that modify gene expression.

In this systemic conception, stemness more precisely appears as a function maintained by the system. At any time, a certain amount of cells are in the stem state, so that the system can "achieve a controlled size, maintain itself homeostatically, and regenerate when necessary" (Lander 2009, 70; see also Wolkenhauer 2001). Such attribution of functions is not specific to the systemic conception of stemness. In all conceptions, stemness is a function that allows a system to develop, maintain itself, and regenerate. This is particularly true for the cancer stem cell theory. Accordingly, it is important to clarify the notion of function and the notion of "systemic property" in the case of stem cells.

Stemness as a Non-dispositional Systemic Function

When biologists say stemness is a function of a system, what do they mean by function? Two notions of function currently coexist: the etiological and the systemic. According to the etiological definition, to attribute a function to an entity is to explain the presence of the concerned trait by some selective advantage that would have allowed it to be selected for in the past (Wright 1973; see also Neander 1991, who suggests to define functions as selected effects). Thus, according to the etiological concept of function, to say that the function of the heart is beating is to explain the presence of the heart (and its beat) by the evolutionary history that led to the selection of this trait. In contrast, the systemic definition of function accounts for the contribution of functions to the explanation of the capacities of the system to which they belong (Cummins 1975). Thus, according to the systemic concept of function, to say that the function of the heart is beating is to say that if the blood circulates, it is, among other things, by virtue of the fact that the heart beats.

When referring to stemness as a function of a system, biologists clearly mean the systemic definition of function: stemness helps explain the capacities of the system (such as the capacity to "achieve a controlled size, maintain itself homeostatically, and regenerate when necessary," mentioned by Lander 2009). However, the idea that stemness is a function of a system slightly differs from Robert Cummins's account of systemic functions. First, describing stemness as a systemic function is not specific to the "systemic conception" of stem cells. What is specific to the idea that stemness is a systemic property is the ascription of the function, or the regulation of the function, to the system. Second, Cummins describes the attribution of function as an attribution of disposition, whereas a systemic property is not a disposition.

Cummins defines the attribution of function as follows:

> x functions as a ϕ in s (or: the function of x in s is to ϕ) relative to an analytical account A of s's capacity to ψ just in case x is capable of ϕ-ing in s and A appropriately and adequately accounts for s's capacity to ψ by, in part, appealing to the capacity of x to ϕ in s. (Cummins 1975, 762)

Translated to stemness in CSCs, the systemic definition of function stipulates that to attribute the stemness function (ϕ) to CSCs (x) in cancers (s), the function must be attributed relative to an explanation A of the system's capacity s (here, the capacity of the cancer) to do something ψ (here, to maintain, develop, and propagate). We can say that stemness (the ability to self-renew and to differentiate, *φ*) of the CSCs *(x)* is their function in cancers *(s)* because the CSC theory's explanation *(A)* of the ability of cancer *(s)* to develop, maintain, and propagate *(ψ)* appeals to the capacity of CSCs *(x)* to self-renew and differentiate *(φ)* in cancers *(s)*. This definition applies to CSCs independent of their identity, that is, independent of whether stemness is a categorical property, a dispositional property, a relational property, or a systemic property. Whatever the kind of property stemness is, the CSC theory explains the development of cancers by appealing to the stemness capacity of CSCs. Consequently, Cummins's account of systemic functions appears unable to grasp the singularity of stemness as a systemic property.

Even worse, as stated by Cummins, a systemic definition of function would rather characterize stemness as a disposition:

> Function-ascribing statements imply disposition statements; to attribute a function to something is, in part, to attribute a disposition to it. If the function of *x* in *s* to *φ*, then *x* has a disposition to *φ* in *s*. (Cummins 1975, 758)

However, this contradiction (ascribing a systemic function would be ascribing a disposition, not a systemic property) can easily be overcome. First, the idea that systemic functions are dispositions comes from the fact that all examples used to study functions are stable entities that are not interchangeable. Examples conventionally used to examine the systemic concept of functions, such as the contractile vacuole of the freshwater vertebrate or the protozoan heart, do not lend themselves to the distinction between assigning a function and assigning a disposition. But there is no reason or argument to justify the claim that ascribing a function is ascribing a disposition, leaving open the opportunity to frame a non-dispositional systemic concept of function. Second, for Cummins, to attribute a disposition is to assert a regularity similar to a law.

> To attribute a disposition *d* to an object *a* is to assert that the behavior of *a* is subject to (exhibits or would exhibit) a certain lawlike regularity. (Cummins 1975, 758)

This is not specific to dispositional properties—it can apply to systemic properties, too. Framing a non-dispositional concept of systemic function would not disrupt the assertion of lawlike regularity. In the CSC theory, the stemness attributed to CSCs and, consequently, the comportment of CSCs are stable and subject to lawlike regularity (see Chapter 2). The fact that CSCs are substitutable and that cancer cells can become CSCs at any time does not change the fact that the cancer cells that exhibit stemness at any time always perform the same function.

Thus, the systemic conception of stemness defines stemness as a non-dispositional systemic function. In other words, stemness is the function of the cells that endorse the CSC identity at any time in a given cancer. Fagan would describe it as follows: "Cell development is conceived as a continuum, different positions on which correspond to different functional roles that cells can occupy. Cells occupy functional roles in virtue of being in particular states" (Fagan 2013a, 72). To better understand what a non-dispositional systemic function is and how it differs from a categorical property or a dispositional property, I suggest the following metaphor. In team sports, such as soccer, there are some field positions to which specific functions are ascribed. For example, the main function of the goalkeeper is to stop the ball from crossing the goal line, whereas the main function of a forward is to advance the ball across the goal line. These functions are examples of non-dispositional systemic functions. They are systemic functions because of their contribution to the explanation of the capacities of the system to which they belong. They are non-dispositional systemic functions because one player can change his or her function and one particular function can be sequentially attributed to different players.

The coach chooses a tactical plan in which each field position has a specific function. This function is a systemic function because the explanation of the capabilities inherent in the game's system adopted by a team calls on activities specific to each position: if players do not perform their functions, the game system will be dysfunctional and will collapse. As such, systemic functions ascribed to the players according to their positions are not dispositions. First, a disposition is intrinsic. It is attributed to the player, not to the field position. Soccer players can change field positions during the game or during their careers. In so doing, they change their functions, whereas their dispositions do not change. Second, substitute players can successively perform the same function as any member in the starting lineup in the course of a match. In contrast with the heart, which is the only entity that can perform its beating function in a given organism, players are interchangeable. In other words, in soccer, attributing a systemic function is not equivalent to attributing a disposition, as it is in the case of the heart.

This example illustrates what I mean by "systemic property" and "non-dispositional systemic function." Two features are central:

1. Reversibility, from the entities standpoint: soccer players are interchangeable in the exercise of the "non-dispositional systemic function" of the different field positions.
2. Robustness, from the function standpoint: in a soccer team, there are always 11 players occupying 11 determined functions; otherwise, the system is dysfunctional.

Thus, for stem cells and, by inclusion, for CSCs, the systemic conception of stemness defines stemness as a function, in a system, imputable to any cell of the system at any time. By extension, at any given time for a given cancer, there is a stock of CSCs that perform the stemness function. It also means that the cells that constitute this stock would be variable from time t_1 to t_n.

Consequences for Therapies

This systemic conception of stemness has problematic consequences for therapeutic strategies against cancers. Neither CSC targeting nor niche targeting can guarantee the success of the cure. The uncertain result of such therapies is amplified by the heterogeneity of the factors involved in stemness acquisition and by the lack of data. Of the two types of factors potentially involved (stochastic and population level), only population-level regulation can be targeted. While the current lack of knowledge about such regulation makes such targeting hypothetical, it should be possible, as in the case of the niche hypothesis, to intervene on some key molecular receptors to prevent or inhibit the induction of stemness by extrinsic factors. Nevertheless, the efficiency of such a therapeutic strategy would depend on the possibility, and probability, of stemness induction by intrinsic stochastic events: the higher the probability of dedifferentiation through stochastic events, the less efficient this therapeutic strategy would be. So, how can we eliminate stemness in cancers if it is a systemic property? To begin to answer this question, we must return to systems biology and the landscape metaphor.

System biology describes cell states as a landscape wherein some states (those located in the valleys) are attractors. Among attractors, stability depends on the depth of the valley. As a consequence, the probability that a cell dedifferentiates to the stem state will depend on the state in which the cell is. A mature red blood cell, for example, is unable to return to the stem state because it has no nucleus. One can legitimately hypothesize that other differentiation states, particularly postmitotic states of differentiation, might also represent highly stable states with a very poor probability of dedifferentiation. The fact that the efficiency of reprogramming technics decreases with cell differentiation supports this idea (see Chapter 6). Pushing the cancer cells to such differentiated states, with a differentiation therapy, could then cure the cancer. More precisely, if stemness is a systemic property, then the efficiency of a differentiation therapy will be inversely equal to the probability that the differentiated cells can dedifferentiate.

Most of the proponents of the CSC theory focus either on the CSC-targeting strategy or on the niche-targeting strategy. However, a few references to differentiation therapies deserve to be highlighted.

From the 1970s to the 1980s, Barry Pierce, who framed an early version of the CSC theory (see Chapter 3), suggested that differentiation therapies could cure cancers. In 2004, Stewart Sell, who worked with Pierce, wrote,

> If the malignant cells of cancers are cancer stem cells, then
> it should be possible to treat cancers by inducing differenti-
> ation of the stem cells, i.e., differentiation therapy.... If
> tumor cells can be forced to differentiate and to cease pro-
> liferation, then their malignant potential will be controlled.
> (Sell 2004, 18)

In practice, very few differentiation therapies have been successfully developed. The most famous ones are retinoic acid (vitamin A) and arsenic trioxide (As_2O_3). Both are differentiating agents used in the treatment of acute promyelocytic leukemia (APL), a leukemia characterized by a translocation between chromosomes 15 and 17 that results in the production of PML/RARα, a fusion protein that causes an arrest of maturation in myeloid cell differentiation at the promyelocytic progenitor stage (Wang and Chen 2008). One retinoic acid, called all-*trans*-retinoic acid (ATRA), showed encouraging results in APL (Tallman et al. 2002; Camacho 2003), and Benito Campos and colleagues, at the University of Heidelberg in Germany, also showed that ATRA targets the CSC population of gliomas, "abrogating major malignancy-related properties of these cells and inducing long-term antiangiogenic, antimigratory, antitumorigenic, proapoptotic, and therapy-sensitizing effects" (Campos et al. 2010, 2716). Retinoic acid and arsenic trioxide restore normal differentiation in APL by inducing PML/RARα degradation. However, separately, they are insufficient to permanently cure the disease. Hugues de Thé's team, at the Hospital Saint Louis in Paris, investigated the mechanisms of retinoic acid and arsenic trioxide therapies and arrived at the conclusion that they must be used in conjunction to both induce differentiation of the cancer cells and prevent self-renewal of the leukemic stem cells (Nasr et al. 2008; Rice and de Thé 2014). Indeed, they observed that restoration of differentiation with retinoic acid was insufficient in APL with another translocation that resulted in the production of another kind of fusion protein called PLZF/RARA, because it failed to deplete the leukemia from CSCs (Nasr et al. 2008). In my lab in Gustave Roussy Hospital, we also observed that hypomethylating agents such as decitabine could restore differentiation without changing the mutation allele burden in the patient—that is, without diminishing the proportion of cancer cells in the blood of the patient (Merlevede et al., *in press*). Thus, induction of differentiation might not be sufficient if CSCs can still self-renew.

Other data also support the differentiation therapeutic strategy. For example, when developing a breast CSC model, Eric Lander's team induced the CSC state through EMT to screen breast CSC-targeting drugs. Using this system, the team identified a differentiation therapy (salinomycin) as the most efficient drug against cells with CSCs' features (Gupta et al. 2009).

Supporting this result, Chia-Ying Lin's team at the University of Michigan tested bone morphogenetic protein-2 (BMP-2), a growth factor that plays a role in inducing the formation of cartilage and bone, and in cell proliferation, apoptosis, and differentiation, on the CSCs of a osteosarcoma cell line (the OS99–1 cell line). Their results suggest that BMP-2 can activate osteogenic differentiation, depleting the CSC population (Wang et al. 2011).

Jörg Glatzle's team, at the University of Tübingen in Germany, also tested a differentiation therapy to fight against the CSCs of gastric cancers. They targeted a metabolic enzyme, the phosphoglycerate kinase 1 (PGK1), known to have an important role in DNA replication and repair, to be involved in many cancers and, more precisely, to be involved in invasion and dissemination of gastric cancers cells. Their results show that in addition to inducing differentiation, anti-PGK1 greatly reduces the invasive potential of gastric cancer cells (Zieker et al. 2013).

These studies highlight the potential usefulness and feasibility of differentiation therapies (for a review of other potential differentiation therapies, see Pattabiraman and Weinberg 2014). An advantage of differentiation therapy is that it could work regardless of the kind of property stemness is.

If stemness is a categorical property or a disposition, then differentiating the CSCs would be equivalent to eliminating them. However, the case of imatinib reminds us that, to work, differentiation therapies must target all the cancer stem cells to be effective. Imatinib mesylate induces cell differentiation, which results in cell death within a few days. The drug achieved great results in CML, but patients relapse when the therapy is interrupted, suggesting that it fails to differentiate the most undifferentiated cancer cells (Goldman 2005; Sell 2006).

If stemness is an extrinsic property that cancer non-stem cells can acquire, Nemanja Marjanovic, Robert Weinberg, and Christine Chaffer (2013) conclude that differentiation therapies are doomed to fail because non-CSCs could permanently turn back to the CSC state. This is true if the differentiation state induced by the therapies is reversible, but system biologists suggest that there might be some highly stable differentiated states that would prevent such dedifferentiation from occurring. As a consequence, differentiation therapies could succeed in curing cancers. However, to achieve their aim, the differentiation therapies must (1) target all the undifferentiated cells and (2) induce a highly stable differentiated state to ensure that the remaining cancer cells have the lowest possible chance of dedifferentiation.

CONCLUSION

If stemness is an extrinsic property that cancer non-stem cells can acquire, the conventional CSC-targeting strategy suggested by the CSC theory may be called into question. According to this strategy, eliminating all the CSCs is necessary and sufficient to cure cancers. But this chapter has shown that if stemness is an extrinsic property, then the eradication of all the CSCs might be necessary, but it will not be sufficient to cure cancers. It is not sufficient because after elimination of all the CSCs, any cancer non-stem cell can still acquire stemness and initiate a relapse. If dedifferentiation is a plausible event in cancers, then either all the cancer cells (including the CSCs) must be eliminated, or dedifferentiation must be blocked and CSCs eliminated.

The data currently available in favor of the idea that stemness is an extrinsic property indicate that two conceptions of the stemness property should be distinguished. Either stemness is specifically induced by the niche, and it is an "extrinsic relational property," or it can be induced by some heterogeneous intrinsic and extrinsic mechanisms at the population level, and it is a "systemic property." The two conceptions are not necessarily mutually exclusive. They can both be used for describing the same cancer or different types of cancers. Niches may have an important role in some cancers and not in others. Thus, determining which factors are at stake in a given cancer is of major importance for optimizing therapeutic strategies. If stemness is a niche-induced property, then niche-targeting therapies might cure cancers. If stemness is a systemic property, cell-cell relation-targeting strategies could cure cancers only in the absence of any other kind of dedifferentiation. The success of such a strategy would be inversely related to the probability of dedifferentiation by factors not targeted by the therapy.

Most of the efforts made under the framework of the CSC theory are focused on either targeting the CSCs or targeting the niche. However, if stemness is a systemic property, both of these therapeutic strategies might fail to bring the expected results. In this case, differentiation therapies might represent a better solution to combat CSCs.

CONCLUSION

       This book provides a critical examination of the concept of cancer stem cell, an old idea that has met with increasing success since the early 2000s. CSCs, a subpopulation of cancer cells with stemness, are at the heart of a theory of cancer development, elaborated around three main theses: (1) that cancers develop and spread exclusively from the subpopulation of CSCs, (2) that CSCs are resistant to therapies and responsible for relapses, and (3) that "in order to cure cancer, it is necessary and sufficient to kill the CSCs" (Reya et al. 2001, 110).

       To understand the true therapeutic potential that investigations of CSCs hold, we must first comprehend the foundations on which the science rests. My task has been to deconstruct both the concept of CSC and the CSC theory of cancer. This task, realized through a detailed philosophical investigation of both historical and current research on cancer stem cells, has unearthed several areas of dissonance within the CSC theory of cancer that all stem from a single problem: stemness identity. Throughout this book, I have highlighted the following:

1. the concept of stem cell is highly ambiguous;
2. the ambiguity of the concept of stem cell concerns the nature of stemness more than its definition: we do not know if, and in which tissues/cancers, stemness is a categorical, a dispositional, a relational, or a systemic property;
3. determining the stemness identity is of paramount importance for cancer treatment because different therapeutic strategies (CSC targeting, niche targeting, differentiation therapies) will have a different outcome depending on the CSCs' identity.

THE CONCEPT OF STEM CELL IS HIGHLY AMBIGUOUS

The ambiguity of the stem cell and CSC concepts is due to several factors. First, both concepts are ambiguous as to their referent (which cells are stem cells/cancer stem cells) and as to their definition. This conclusion was drawn from a detailed historical analysis beginning with cell theory in the nineteenth century, wherein we found the origins of the central hypothesis of the CSC theory—namely, that cancers originate from stem cells. Throughout this history, both "stem cell" and "CSC" were used to refer to a variety of cells. Second, self-renewal and differentiation, the two properties that constitute stemness, are neither homogeneous among different types of stem cells nor specific to stem cells. Together, these difficulties raise questions about the nature of the stem cell category and, more generally, about the identity of stem cells.

The Ambiguity of the Concept of Stem Cell Concerns
the Nature of Stemness More Than Its Definition

Within the classical view of stem cells, the property of stemness is considered an intrinsic property—that is, stem cells are a natural kind or type that are inherently endowed with the ability to self-renew and to differentiate. This view of stem cells and stemness as an intrinsic property is widely accepted throughout biology. However, there have been calls from stem cell biologists to recognize stemness as an extrinsic property that non-stem cells can acquire, arguing that the concept of stem cell should be understood as referring to a "state" (Dov Zipori) or to a "function" (Helen Blau) rather than to an "entity." This move to recognize stemness as an extrinsic property, conferred upon stem cells rather than inherent to them, pushes stem cells away from being a determined cell type and toward stem cells being a cell state, characterized by reversibility and transiency.

My philosophical analysis of the conception of stemness has shown that there are four distinct ways in which stem cell biologists conceive of and use stemness. These four ways of understanding stemness are implicitly engrained within stem cell biology and guide the research of biologists so thoroughly that it is worth taking the time to explicate their definitions.

If stemness is intrinsic to the cell, then it may be either a "categorical property" or a "dispositional property." A categorical property is an essential and intrinsic property that is stable and specific to stem cells, whereas a dispositional property is one whose expression depends on activation by external stimuli, such as the niche. Stemness as a dispositional property reflects a weak interpretation of the niche hypothesis, according to which the stem cell niche regulates stemness activity.

If stemness is extrinsic to the cell, then it may be either an extrinsic relational property or a systemic property. An extrinsic relational property is characterized by a strong interpretation of the niche hypothesis, according to which the niche induces stemness, thus producing new stem cells. In this case, the relationship of the cell with the niche determines the identity of the cell. Meanwhile, a systemic property is one in which the properties of stemness (self-renewal and differentiation) are a function of the system (i.e., conferred and maintained by the system), rather than of the individual cells. The boundaries of the "system" for the CSC theory include at least all the cancer cells of a given cancer. The key idea of the systemic conception is that stemness is maintained at the system level. At any time, some cells of the system are in the stem state, thus ensuring the system's functionality. In this view, the factors that can induce the stem state can be of various kinds, intrinsic or extrinsic.

Table C.1. CSCs' identity and the effectiveness of therapeutic strategies against cancers.

	CSCs-targeting therapies	Combined therapies	Niche-targeting therapies	Differentiation therapies
Categorical property	Yes	Yes	No	Yes
Dispositional property	Yes	Yes	Yes	Yes
Relational-extrinsic property	No	Yes	Yes	Likely
Systemic property	No	Yes	No	Likely

Data are currently insufficient to determine the actual identity of stem cells and cancer stem cells. More studies are required to settle the issue. The question is then whether and why biologists should engage in such research. This book provides an answer for the case of CSCs in oncology: stemness identity matters for oncology because the success of different therapies relies on the identity of CSCs.

Determining the Identity of Stem Cells Is of Paramount Importance for Cancer Treatments

If the CSC theory of cancer holds any value for treating and curing cancers, then understanding which of the four conceptions of stemness most accurately defines stem cells holds incredible therapeutic value. Table C.1 summarizes the different possible relationships between therapeutic strategies and the four possible stemness properties. The underlying assumption of this table, and of therapeutic strategies more generally, is that the elimination of CSCs is *necessary* to cure cancers—an assumption that needs to be tested by determining whether the CSC theory is valid for each type of cancer.

Table C.1 makes it clear that each conception of stemness has implications for therapeutic strategies. For instance, the strategy of targeting CSCs can only be said to be necessary and sufficient to cure cancer if stemness is a categorical or dispositional property, and its efficiency will correlate with its actual ability to kill all the CSCs. If stemness is an extrinsic relational or a systemic property, then a cancer non-stem cell could acquire stemness and initiate relapse at any time. Consequently, CSC targeting might not be sufficient to ensure recovery. The risk of relapse following this kind of therapy will correlate with the probability that non-CSCs acquire stemness in this type of cancer. Meanwhile, if stemness depends on the niche, either as a dispositional or as an extrinsic relational property, then another therapeutic strategy is possible: targeting the niche-cell interactions. Such a therapeutic strategy, however, would only be effective and efficient if the targeted interactions are necessary for the expression of stemness.

Only one therapeutic strategy could work irrespective of the nature of stemness: the addition of traditional cancer therapies to CSC-targeting therapies, resulting in a cumulative strategy for cancer therapy. Such cumulative therapeutic strategies are often suggested in the CSC literature. However, deploying them raises a number of issues. In particular, cumulative therapeutic strategies raise the issue of the biological cost for the patient (e.g., the multiplication of side effects) and of the financial cost for the health care systems.

Another therapeutic strategy, which is not specific to CSCs, appears likely to be successful irrespective of the nature of stemness: differentiation therapies—i.e., therapeutic interventions in which cancer cells are induced to differentiate. Inducing differentiation of all the cancer cells would result in the elimination of the CSCs. Therefore, if stemness is a categorical or dispositional property, then one would expect the same results from a differentiation therapy as from a CSC-targeting therapy. Thus, differentiation therapies may be sufficient to cure cancers, provided that they also prevent CSCs from self-renewing. The situation is less clear if stemness is a systemic or an extrinsic relational property because differentiated cells could still reacquire stemness. However, one can reasonably assume that some states of differentiation are not readily reversible and possibly irreversible. If differentiation therapies can induce such highly stable states of differentiation, then, at the very least, there would be a very low probability of relapse.

Based on the observations that (1) CSC-targeting strategies face many difficulties, (2) it is unclear whether CSC targeting is sufficient to cure cancers, (3) knowledge is crucially lacking to target niche-cell interactions, and (4) it is also unknown whether such targeting therapies would be sufficient to cure cancers, differentiation therapeutic strategies may be the most reasonable therapeutic strategy in the current state of knowledge. This observation contrasts deeply with the poor attention that differentiation therapies receive from the CSC research community.

This exploration of the relationship between the four conceptions of stemness and therapeutic strategies has a surprising consequence: all

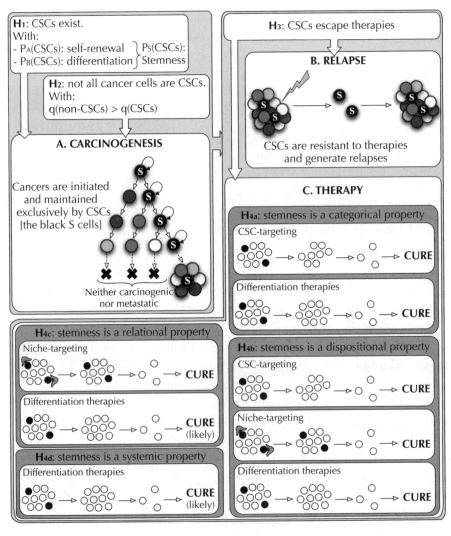

Figure C.1. The CSC therapies with the hypotheses on stemness identity. (Illustration © 2016 by Lucie Laplane.)

the premises of the CSC theory could be true and the therapeutic model could prove to be false. This is because the current therapeutic strategies based on CSCs are based on two questions: Do CSCs exist? If yes, then their elimination is assumed to be sufficient for cancer therapy. And, how do we define stem cells? The common answer is that we define stem cells by the presence of stemness, which is composed of two properties: self-renewal and differentiation. However, this text has shown that these two questions are not sufficient when it comes to affecting a cure for cancer—a question about the nature of stemness needs to be added to the premises of the CSC theory in order for CSC therapies to truly be effective (see Figure C.1).

PERSPECTIVES

In addition to therapeutic implications, a number of philosophical and biological questions arise from the realization that stemness can be conceived of in four ways:

1. Does stem cell identity fall into one or several of the conceptions of stemness?
2. Are they all equally plausible?
3. Are they empirically testable?
4. What are the other consequences of stem cell identity?

The first question can only be resolved through investigations along multiples lines of inquiry, including better understanding of dedifferentiation and the cellular niche in different contexts. A large number of data indicate that dedifferentiation is possible. Similarly, existence of cellular niches, in which stem cells are located and with which stem cells interact, is hardly debatable. Therefore, these issues must be addressed through reference to the following questions:

1. To what extent is dedifferentiation likely to occur? The answer to this question is susceptible to vary from one species to another, from one tissue to another, and from one developmental context to another.
2. What is the role of the niche? Is it only involved in the regulation of stemness activity or is it also capable of inducing stemness in

non-stem cells? To what extent are the interactions with the niche necessary for stemness activity? Here again, the answer is likely to vary depending on the species, tissues, and developmental contexts.

For example, epithelial tissues such as the colon seem prone to dedifferentiation and stemness acquisition (Donati and Watt 2015; Tetteh, Farin, and Clevers 2015), whereas no data have ever reported stemness acquisition in the hematopoietic system. Given that the niche seems to play a determinant role in both stem cell types (Barker 2013; Boulais and Frenette 2015), stemness could be an extrinsic relational property in colon stem cells and a dispositional property in hematopoietic stem cells.

The identity of stemness can also be different in cancers compared to normal tissues. Some cancers might increase the cellular plasticity and the probability of stemness acquisition, for example, by increasing the probability of epithelial-mesenchymal transitions (Theise and Wilmut 2003; Chaffer et al. 2011). Mutations can also modify the properties of cancer cells. For example, some mutations induce constitutive activation pathways that make the cell independent from some extrinsic signaling usually required for CSC proliferation (Hanahan and Weinberg 2011; Feitelson et al. 2015). Thus, an open question is whether such mutations can transform a dispositional property into a categorical property. One last consideration is that some cancer mutations also seem able to induce stemness in non-stem cells, such as *MOZ-TIF2* (Huntly et al. 2004) and *MLL-ENL* (Cozzio et al. 2003) in AML, which means that new populations of CSCs can be generated from non-stem cells even in tissues with a strict unidirectional hierarchy.

The question of whether each of the conceptions of stemness is equally plausible can be addressed from two perspectives: metaphysical and methodological. The metaphysical question is, in light of the data, is it more likely that stemness is a categorical, dispositional, extrinsic relational, or systemic property? The methodological question is, in the face of uncertainty, which conception is preferable? With regard to the metaphysical view, for some stem cells, data suggest that some conceptions are more plausible than others. For example, stemness seems more likely to be a dispositional property in hematopoietic stem cells and a relational property in colon stem cells. However, for most stem cells, the data are not clear enough, and further investigations are necessary to clarify the status of stemness identity. In such cases, I maintain an agnostic position—stemness identity cannot be settled until more experiments are done. In contrast, in the case of the methodological perspective, adopting one view over the others has consequences that are not strictly equivalent in terms of research requirements. Thus, if we *must* adopt one of these conceptions without enough data to support this choice, then we should adopt the most exigent one. For cancer therapies, this is the systemic conception. Indeed, adopting this conception leads us to consider targeting CSCs and extrinsic factors as possibly insufficient for curing cancers. In addition, adopting the systemic conception encourages us to consider all possible factors, making it very open in terms of research programs. Despite this methodological preference, I support agnosticism with regard to the identity of stem cells—one cannot and must not conclude on the exact nature of stemness. On the contrary, the current state of uncertainty and the need to assess the nature of stemness must be acknowledged.

With regard to the third question of whether or not the four conceptions of stemness are testable, the answer is: they are, in principle. It would suffice to show that a cancer non-stem cell can acquire stemness (i.e., becomes a CSC) and to quantify the probability of such an event in cancers, to end the alternative between intrinsic and extrinsic, or *entity* and *state*. If the probability is so low that dedifferentiation could be considered an exception, then therapeutic strategies can be evaluated in the light of the *entity* view. If the probability is high enough to represent a risk, then the *state* view should be endorsed. Thus, the issue becomes assessing whether a niche is necessary for cells to act as stem cells. This would allow us to know whether stemness is a categorical property or a dispositional property, in one case, and if it is a systemic property or an extrinsic relational property, in the other.

New technologies now offer us tools to begin testing stem cell identity. Consider, for example, the multicolor clonal cell tracking enabled by the Brainbow technology (e.g., Livet et al. 2007). It allows cells to coexpress stochastically different fluorescent proteins, which show up as different colors on confocal microscopes. Dozens of different colors can be generated by multiplying the Brainbow tandems, distinguishing many cells and their progeny. In addition to Brainbow technology, Cre-Lox recombination technology make it possible to activate multicolor fluorescent labeling at specific times, in targeted cell populations. This technology opens many doors for the study of cell lineages. For instance, it is possible to activate Brainbow tagging in different cell types at different stages of differentiation to observe the behavior of the cellular progeny—for example, their ability to dedifferentiate in particular contexts.

Cellular barcoding is another new technology that allows us to track a single cell's progeny (e.g., Naik et al. 2013; Naik, Schumacher, and Perié 2014; Perié et al. 2014). Cellular barcoding consists of extracting cells of interests, tagging each cell through transfections of a unique small DNA sequence, and transplanting the tagged cells into mice. As the barcode is transmitted through cell division, it is possible to assess the contribution of each cell in the production of a tissue, like hematopoiesis. Such a technology could be used to assess the ability of non-stem cells to acquire stemness by tracking their ability to generate new populations of heterogeneous cells over a long period of time.

Furthermore, the development of anti-CSC therapies could also be seen as tests, in the long term, of the categorical and dispositional conceptions. Indeed, if such therapies were developed and consistently resulted in the complete and definitive recovery of patients, it would provide a retrospective demonstration of the categorical or dispositional nature of stemness. Similarly, niche-targeting therapies (or CSC-niche relation-targeting therapies) might bring a retrospective demonstration of the dispositional or extrinsic relational nature of stemness.

Finally, regarding the last question of the implications of the four conceptions outside of cancers, one can quite easily guess important consequences for regenerative medicine. The use of stem cells as therapeutic tools is necessarily affected by the nature of stemness. For example, if stemness is a dispositional or relational property, then stemness cannot be attributed to the stem cells alone but must be attributed to the niche-cell complex. This raises the issue of what needs to be transplanted and where, for regenerative therapies to succeed. Can we transplant stem cells alone, or do they need their supportive cells? Do we need to transplant the cells directly into the right niche? We might also have to take into account the external environment, such as light (circadian circle). Light, for example, regulates the production of SDF1 by the hematopoietic stem cell (HSC) niche. As SDF1 activates *CXCR4,* a gene that plays a major role in the anchoring of HSCs to the niche and in the regulation of quiescence and self-renewal of the HSCs, it raises the question of when we should transplant the cells—during the day or at night?

Regardless of what direction future questions about the nature of stemness take research, the imperative is clear: a better understanding of the nature of stemness has benefits to the research communities that investigate stem cells and CSCs, benefits to the biomedical industry for the treatment of cancers and regenerative therapies, and benefits to the public more broadly for their therapeutic values.

NOTES

INTRODUCTION

1. http://www.ligue-cancer.net/article/9523_le-cancer-en-france#.Ue6
   iPFPOuRt.
2. The National Cancer Act of 1971, Pub. L. No. 92–218, 85 Stat. 778 (1971).
3. Mount Sinai Press Release, "Mount Sinai Researchers Unveil New
   Chemotherapy-Resistant Cancer Stem Cell," September 10, 2012. See
   also Wicha (2014).
4. For more on this point, see Laplane (2011).

1. CANCER STEM CELLS' TRIUMPH

1. *Cell Stem Cell, Oncogenesis, Tumor Biology, Cancer Cell, Carcinogenesis, Current Opinion in Biotechnology, Laboratory Investigation, Médecine/Science, The
   Journal of Clinical Investigation, Neurobiology of Disease, Cell Cycle,* and *Cancer
   Cell* are just a few examples.
2. For example, in July 2014, *Web of Science* reported 3,539 citations of Reya
   et al. (2001). It is, by far, the most cited article of the November 1 issue,
   before Zhou et al. (2001) (1,097 citations).
3. Bayard Clarkson is an oncologist specializing in leukemia. He has been
   involved with the AACR, for which he served as president in 1980 and
   treasurer for 15 years. His team works on the differences between normal
   and cancer stem cells, particularly in chronic myelogenous leukemia.
4. Crimson Foundation, course PABSELASCRT08; Cold Spring Harbor
   Laboratory Courses, "Mouse Development, Stem Cells & Cancer," June
   5–25, 2013.
5. Interview with Dominique Bonnet, Mandelieu, April 5, 2009.
6. Interview with Nicole Gross, Lausanne, July 29, 2009.

7. Interview with Marie-Pierre Junier, Mandelieu, April 4, 2009.

8. During research conducted on July 22, 2015, I was able to find this figure in other publications (Das, Srikanth, and Kessler 2008; RicciVitiani et al. 2008; Diehn et al. 2009; Watt and Driskell 2010; Matos Rojas, Bertholdo, and Castillo 2012; McAuliffe et al. 2012; Succony and Janes 2014); in the news section of the "Radiological Society of North America," in the article "stem cells and cancer" in the journal *Stem Cell Lines* (published three times per year) addressed to the Harvard Stem Cell Institute's fellows, in the "Fiche info" of the stem cells European portal; in a blog post of the Stem Cells Research blog; in the "Cancer Stem Cell" entry of Wikipedia; and in the stem cells startup Currin's weekly newsletter, in the New York–based biopharmaceutical company Stemline's website, in the Stockholm-based biopharmaceutical company Cormorant, and in the biotechnology company YZYBiopharma's website. I was also able to find this figure translated in Italian, Polish, Swedish, Danish, Japanese, Chinese, German, Spanish, Persian, Korean, Arabic, Tagalog, Tamul, and Thai.

9. OncoMed was also launched by Muhammad Al-Hajj and Max Wicha. Those four researchers are the ones who first observed CSCs in a solid cancer, a breast cancer (Al-Hajj et al. 2003).

10. http://www.verastem.com/.

11. http://www.stemline.com/csc.asp.

12. http://kalobios.com/kb004.

13. See the Technology Development & Commercialization website of the Toronto University Health Network (UHN). Quotation from the July 15, 2009, News titled "UHN, OICR and Pfizer Colorectal Cancer Collaboration."

2. THE CSC THEORY

1. On whether the CSC theory really is a scientific theory, see Laplane (2014).

2. Every biology textbook defines stem cells through these two properties. For example, go to the National Institutes of Health website, which provides a useful glossary.

3. Those are just two examples picked among many others. See, for other recent examples, Cetin and Topcul (2012), Duggal et al. (2013), Wang et al. (2013), and Flemming (2015).

3. TERATOCARCINOMAS AND EMBRYONIC STEM CELLS

1. See Dupont and Schmitt (2004), particularly the first part, which deals with embryonic layers.

2. It is worth correcting a widespread reference mistake in respect of Julius Cohnheim: the articles "Ueber Entzuendung und Eiterung" published in 1867 and "Congenitales, quergestreiftes Muskelsarkom der Nieren" published in 1875 are often cited as a reference of Julius Cohnheim's theory. However, the first one, which we might translate as "On inflammation and suppuration," does not deal with the embryonic rest theory. Therefore, only the second one could be cited as a reference for the embryonic rest theory. See Cohnheim (1867, 1875).

3. The enamel organ has cells that initiate dental development. It is composed of three layers: inner enamel epithelium, outer enamel epithelium, and stratum intermedium.

4. For the embryonic rest hypothesis in the jaw adamantinomas, see Magitot (1860), Koelliker (1868, 1882), Legros and Magitot (1873, 1879), and Malassez (1885a, 1885b). For the embryonic rest hypothesis in the pituitary adamantinomas, see Erdheim (1903, 1904), Lewis (1910), Duffy (1920), Critchley and Ironside (1926), Atwell (1926), Carmichael (1931), Cushing (1932), Susman (1932), Oliver and Scott (1934), and Drummond (1939). For the embryonic rest hypothesis in the tibial adamantinomas, see Fischer (1913), Richter (1930), Baker and Hawkley (1931), Wolfort and Sloane (1938), Rankin (1939), Zegarelli (1944), Schulenburg (1951), Ewing (1940, chap. 6), Rosai (1969), Lichtenstein (1977), Rosai and Pincus (1982), Ishida et al. (1992), Hazelbag et al. (1993), Jundt et al. (1995), and Jain et al. (2008).

5. Terminology for these cancers has been quite malleable. Teratomas and teratocarcinomas were referred to as "embryomas" in the twentieth century, and a precise distinction between teratoma and teratocarcinoma took a long time to take effect. See Morange (2006) and Damjanov and Andrews (2007).

6. Furthermore, it seems that the trophoblastic version of this hypothesis, which John Beard defended, had a new success since the 1970s, particularly in relation to the CSC theory in 2008. Burleigh, for example, claims that the trophoblastic theory is "remarkably similar" to the cancer stem cell theory. See Gurchot (1975) and Burleigh (2008).

7. For more about why it took so long to isolate human ES cells, see Morange (2006) and Brandt (2012).

8. However, Beatrice Mintz defended the hypothesis that teratocarcinomas could have an embryonic somatic origin. See Mintz, Cronmiller, and Custer (1978).

9. For an image, see *Text-Figure* 1 in Pierce (1974).

1. Philip Fialkow's unexpected death in 1996 was marked with a symposium titled "Clonal Origin of Leukemia—Revisited. A Tribute to Philip J Fialkow" during the annual meeting of the American Society of Hematology and eventually led to publications in the journal *Leukemia,* among which was a historical review: Raskind, Steinmann, and Najfeld (1998). See also the editorial by Najfeld (1998).

2. For the idea that leukemic proliferation is uncontrolled and rapid, see references in Killmann (1965).

3. See fig. 1, p. 434, in Till and McCulloch (1980). This hierarchy has evolved ever since and is still evolving.

4. The first clinical trial happened in 1991. See Blaese et al. (1995).

5. For bone sarcomas, see Gibbs et al. (2005). For prostate cancers, see Collins et al. (2005) and Lawson and Witte (2007). For ovarian cancers, see Szotek et al. (2006). For liver cancers, see Suetsugu et al. (2006) and Chiba et al. (2007). For head-and-neck cancers, see Prince et al. (2007). For pancreatic cancers, see Li et al. (2007). For colorectal cancers, see Dalerba et al. (2007) and O'Brien et al. (2007). For kidney cancers, see Eramo et al. (2008).

5. ORIGIN, STEMNESS, AND STEM CELLS

1. The mouse model only partially explains this result. Morrison's team also transplanted aggressive cells.

2. According to the hypothesis of the mutator phenotype, some cancerous mutations would increase genetic instability by making the cell prone to accumulate more mutations (see Loeb 1991; Loeb and Loeb 2000; Beckman and Loeb 2006). For critics, see Komarova (2004) and Komarova and Wodarz (2004).

3. On the clonal evolution theory and other Darwinian models of cancer cell evolution, see also Cairns (1975), Greaves (2000), Gatenby and Vincent (2003), Merlo et al. (2006), Vineis and Berwick (2006), and Frank (2007).

6. STEM CELL IDENTITY

1. The distinction between totipotent and pluripotent stem cells only applies to those species, such as mammals, for which development requires extraembryonic tissues. Philosopher Christine Hauskeller has pointed out that ethical debates on embryonic stem cells have helped to motivate a rigorous distinction between totipotent and pluripotent stem cells. Previous to these debates, pluripotency was used quite loosely. With eth-

ical concerns on stem cell research, biologists argued in favor of a strict distinction between totipotent and pluripotent stem cells to pinpoint that the embryonic stem cells extracted from the inner cell mass of the embryo and cultured *in vitro* cannot develop into organisms. See Hauskeller (2005).

2. For examples of essentialism about natural kinds, see Ellis (2001). For examples of anti-essentialism, see Mellor (1977) and Dupré (1981).

3. *In situ* hybridization consists of observing the expression of a gene through the labeling of RNA corresponding thereto. The labeling is done using a probe that is a complementary strand to the target sequence (antisense RNA) that will hybridize with the RNA molecule. The probe itself is marked by the digoxygenin-UTP (DIG). Thus, treatment from the anti-DIG antibody coupled to alkaline phosphatase allows one to "reveal" the presence of the target RNA through a purple precipitate generated by the antibody when degraded in the alkaline pH medium. Antibody labeling consists of observing the expression of a gene through the labeling of a protein corresponding thereto. The labeling is done with a monoclonal antibody specific to the protein itself labeled with a fluorescent secondary antibody, observable with a confocal microscope.

4. Note that an analysis of the ES cell line showed considerable heterogeneity between these cells, despite the large number of genes they express (Osafune et al. 2008).

5. For more information, see the Genetics Home Reference website of the National Institutes of Health: http://ghr.nlm.nih.gov/gene/ITGA6.

6. Liu (2001) also gave a similar interpretation.

7. "STAP Retracted," *Nature,* July 2, 2014: http://www.nature.com/news/stap-retracted-1.15488.

7. IF STEMNESS IS A CATEGORICAL OR A DISPOSITIONAL PROPERTY, HOW CAN WE CURE CANCERS?

1. Categoricalists such as David Armstrong consider that dispositions are reducible to some causal bases of the object and therefore that dispositions are reducible to categorical properties. The conception of the stem cell I am trying to depict here differs from this view insofar as it admits the existence of some categorical causal bases but refuses that stemness can be reduced to them. The only possible categoricalist interpretation of stemness here would imply changing the object that bears the causal bases: instead of the stem cell, the causal bases of stemness would belong to the stem cell-niche complex.

2. For example, the International Cancer Microenvironment Society launched, in 2008, a journal specifically dedicated to research on the niche in cancers, called *Cancer Microenvironment.*

3. Other studies brought arguments in favor of CSCs being independent from the niches. See, for example, Lane et al. (2011).
4. Reviewed in Chen, Huang, and Chen (2013).
5. See https://med.stanford.edu/profiles/irving-weissman?tab=research-and-scholarship.
6. http://clinicaltrials.gov/show/NCT01060319.
7. Cellerant Therapeutics, http://www.cellerant.com/.
8. This metaphor is quite accurate for my argument here. However, it should be acknowledged that Quesenberry and his colleagues would rather adopt the systemic conception of stem cells described in Chapter 8. Moreover, they quickly abandoned this metaphor in favor of a continuum metaphor, in opposition to the entity view or to the existence of a stem cell natural kind.

8. IF STEMNESS IS A RELATIONAL OR A SYSTEMIC PROPERTY, HOW CAN WE CURE CANCERS?

1. The idea that dedifferentiation happen more often in cancers is also expressed in Theise and Wilmut (2003).
2. "Being fifty miles south of a burning barn" is an example of a relational property that was initially suggested by Kim (1973, 231). Many other philosophers have used it since: see, for example, Hoffmann-Kolss (2010, 118) and Kistler (2002, 249).

REFERENCES

Abbott, B. L. 2006. ABCG2 (BCRP): A cytoprotectant in normal and malignant stem cells. *Clinical Advances in Hematology and Oncology* 4 (1): 63–72.

Abou-Khalil, R., F. Le Grand, G. Pallafacchina, S. Valable, F.-J. Authier, M. A. Rudnicki, R. K. Gherardi, S. Germain, F. Chretien, A. Sotiropoulos, P. Lafuste, D. Montarras, and B. Chazaud. 2009. Autocrine and paracrine angiopoietin 1/Tie-2 signaling promotes muscle satellite cell self-renewal. *Cell Stem Cell* 5 (3): 298–309.

Abraham, Carolyn. 2006. Meet the A-Team of stem-cell science. *The Globe and Mail*, November 25: http://www.theglobeandmail.com/news/national /meet-the-a-team-of-stem-cell-science/article18178732/?page=all.

Agliano, A., I. Martin-Padura, P. Mancuso, P. Marighetti, C. Rabascio, G. Pruneri, L. D. Shultz, and F. Bertolini. 2008. Human acute leukemia cells injected in NOD/LtSz-scid/IL-2Rgamma null mice generate a faster and more efficient disease compared to other NOD/scid-related strains. *International Journal of Cancer* 123 (9): 2222–2227.

Aguirre-Ghiso, J. A. 2006. The problem of cancer dormancy: Understanding the basic mechanisms and identifying therapeutic opportunities. *Cell Cycle* 5 (16): 1740–1743.

——. 2007. Models, mechanisms and clinical evidence for cancer dormancy. *Nature Reviews Cancer* 7 (11): 834–846.

Akashi, K., X. He, J. Chen, H. Iwasaki, C. Niu, B. Steenhard, J. Zhang, J. Haug, and L. Li. 2003. Transcriptional accessibility for genes of multiple tissues and hematopoietic lineages is hierarchically controlled during early hematopoiesis. *Blood* 101 (2): 383–389.

Al-Assar, O., R. J. Muschel, T. S. Mantoni, W. G. McKenna, and T. B. Brunner. 2009. Radiation response of cancer stem-like cells from established human cell lines after sorting for surface markers. *International Journal of Radiation Oncology Biology Physics* 75 (4): 1216–1225.

Albini, A., E. Cesana, and D. M. Noonan. 2011. Cancer stem cells and the tumor microenvironment: Soloists or choral singers. *Current Pharmaceutical Biotechnology* 12 (2): 171–181.

Al-Hajj, M., M. S. Wicha, A. Benito-Hernandez, S. J. Morrison, and M. F. Clarke. 2003. Prospective identification of tumorigenic breast cancer cells. *Proceedings of the National Academy of Sciences USA* 100 (7): 3983–3988.

Allan, A. L., ed. 2011. *Cancer Stem Cells in Solid Tumors*. New York: Springer.

American Cancer Society. 2012. *Cancer Treatment and Survivorship Facts & Figures 2012–2013*. Atlanta: American Cancer Society.

———. 2014. *Cancer Facts & Figures*. Atlanta: American Cancer Society.

Anderson, D. J., F. H. Gage, and I. L. Weissman. 2001. Can stem cells cross lineage boundaries? *Nature Medicine* 7 (4): 393–395.

Anderson, K., C. Lutz, F. W. van Delft, C. M. Bateman, Y. Guo, S. M. Colman, H. Kempski, A. V. Moorman, I. Titley, J. Swansbury, L. Kearney, T. Enver, and M. Greaves. 2011. Genetic variegation of clonal architecture and propagating cells in leukaemia. *Nature* 469 (7330): 356–361.

Arai, F., A. Hirao, M. Ohmura, H. Sato, S. Matsuoka, K. Takubo, K. Ito, G. Y. Koh and T. Suda. 2004. Tie2/Angiopoietin-1 signaling regulates hematopoietic stem cell quiescence in the bone marrow niche. *Cell* 118 (2): 149–161.

Arai, F., and T. Suda. 2008. Quiescent stem cells in the niche. In *The Stem Cell Research Community Edition*, edited by StemBook: http://www.stembook.org.

Arechaga, J. 1993. On the boundary between development and neoplasia: An interview with Professor G. Barry Pierce. *International Journal of Developmental Biology* 37: 5–16.

Askanazy, M. 1907. Die Teratome nach ihrem bau, ihrem Verlauf, ihrer Genese und in Vergleich zum experimentellen Teratoid. *Verhandlungen der Deutschen Gesellschaft für Pathologie* 1: 39–82.

Astaldi, G., A. Allegri, and C. Mauri. 1947. Experimental investigations of the proliferative activity of erythroblasts in their different stages of maturation. *Experientia* 3 (12): 499–500.

Astaldi, G., and C. Mauri. 1950. New criteria to estimate mitotic activity in bone marrow cells. *Acta Haematologica* 3 (2): 122–122.

———. 1953. Recherches sur l'activité proliférative de l'hémocytoblaste de la leucémie aigue. *Revue belge de pathologie et de médecine expérimentale* 23 (2): 69–82.

Atwell, W. J. 1926. The development of the hypophysis cerebri in man, with special reference to the pars tuberalis. *American Journal of Anatomy* 37 (1): 159–193.

Auffinger, B., A. L. Tobias, Y. Han, G. Lee, D. Guo, M. Dey, M. S. Lesniak, and A. U. Ahmed. 2014. Conversion of differentiated cancer cells into cancer stem-like cells in a glioblastoma model after primary chemotherapy. *Cell Death and Differentiation* 21 (7): 1119–1131.

Bachem, M. G., K. M. Sell, R. Melchior, J. Kropf, T. Eller, and A. M. Gressner. 1993. Tumor necrosis factor alpha (TNF alpha) and transforming growth factor beta 1 (TGF beta 1) stimulate fibronectin synthesis and the transdifferentiation of fat-storing cells in the rat liver into myofibroblasts. *Virchows Archiv B, Cell Pathology Including Molecular Pathology* 63 (2): 123–130.

Back, J., A. Dierich, C. Bronn, P. Kastner, and S. Chan. 2004. PU.1 determines the self-renewal capacity of erythroid progenitor cells. *Blood* 103 (10): 3615–3623.

Bagley, R. G., and B. A. Teicher, eds. 2009. *Stem Cells and Cancer.* New York: Springer.

Baguley, B. C. 2010. Multiple drug resistance mechanisms in cancer. *Molecular Biotechnology* 46 (3): 308–316.

Baker, A. H., and L. M. Hawkley. 1931. A case of primary adamantinoma of the tibia. *British Journal of Surgery* 18 (71): 415–421.

Bao, S., Q. Wu, R. E. McLendon, Y. Hao, Q. Shi, A. B. Hjelmeland, M. W. Dewhirst, D. D. Bigner, and J. N. Rich. 2006. Glioma stem cells promote radioresistance by preferential activation of the DNA damage response. *Nature* 444 (7120): 756–760.

Bapat, S. A. 2007. Evolution of cancer stem cells. *Seminars in Cancer Biology* 17 (3): 204–213.

———, ed. 2008. *Cancer Stem Cells: Identification and Targets.* Hoboken, NJ: Wiley.

Barabe, F., J. A. Kennedy, K. J. Hope, and J. E. Dick. 2007. Modeling the initiation and progression of human acute leukemia in mice. *Science* 316 (5824): 600–604.

Barda-Saad, M., Y. Shav-Tal, A. L. Rozenszajn, M. Cohen, A. Zauberman, A. Karmazyn, R. Parameswaran, H. Schori, H. Ashush, A. Ben-Nun, and D. Zipori. 2002. The mesenchyme expresses T cell receptor mRNAs: Relevance to cell growth control. *Oncogene* 21 (13): 2029–2036.

Barker, N. 2013. Adult intestinal stem cells: critical drivers of epithelial homeostasis and regeneration. *Nature Reviews Molecular Cell Biology* 15 (1): 19–33.

Barnes, D. W., C. E. Ford, S. M. Gray, and J. F. Loutit. 1959. Spontaneous and induced changes in cell populations in heavily irradiated mice. *Progress in Nuclear Energy—Biological Sciences* 6 (2): 1–10.

Barroca, V., B. Lassalle, M. Coureuil, J. P. Louis, F. Le Page, J. Testart, I. Allemand, L. Riou, and P. Fouchet. 2009. Mouse differentiating spermatogonia can generate germinal stem cells in vivo. *Nature Cell Biology* 11 (2): 190–196.

Beard, J. 1900. The morphological continuity of the germ cells in Raja batis. *Anatomischer Anzeiger* 48 (20–21): 465–485.

———. 1902. Embryological aspects and etiology of carcinoma. *Lancet* 159 (4112): 1758–1761.

———. 1905a. The cancer problem. *Lancet* 165 (4249): 281–283.

———. 1905b. The cancer problem and cancer research. *Lancet* 165 (4250): 385–386.

Becker, A. J., E. A. McCulloch, and J. E. Till. 1963. Cytological demonstration of the clonal nature of spleen colonies derived from transplanted mouse marrow cells. *Nature* 197: 452–454.

Beckman, R. A., and L. A. Loeb. 2006. Efficiency of carcinogenesis with and without a mutator mutation. *Proceedings of the National Academy of Sciences USA* 103 (38): 14140–14145.

Bergsagel, D. E., and F. A. Valeriote. 1968. Growth characteristics of a mouse plasma cell tumor. *Cancer Research* 28 (11): 2187–2196.

Bernardos, R. L., L. K. Barthel, J. R. Meyers, and P. A. Raymond. 2007. Late-stage neuronal progenitors in the retina are radial Muller glia that function as retinal stem cells. *Journal of Neuroscience* 27 (26): 7028–7040.

Berridge, M. V., P. M. Herst, and A. S. Tan. 2010. Metabolic flexibility and cell hierarchy in metastatic cancer. *Mitochondrion* 10 (6): 584–588.

Bhatia, R., M. Holtz, N. Niu, R. Gray, D. S. Snyder, C. L. Sawyers, D. A. Arber, M. L. Slovak, and S. J. Forman. 2003. Persistence of malignant hematopoietic progenitors in chronic myelogenous leukemia patients in complete cytogenetic remission following imatinib mesylate treatment. *Blood* 101 (12): 4701–4707.

Bird, A. 1998. Dispositions and antidotes. *Philosophical Quarterly* 48 (191): 227–234.

Bird, A., and E. Tobin. 2012. Natural kind. In *The Stanford Encyclopedia of Philosophy,* edited by E. N. Zalta: http://plato.stanford.edu/archives/win2012/entries/natural-kinds/.

Blackstock, A. M., and O. M. Garson. 1974. Direct evidence for involvement of erythroid cells in acute myeloblastic leukaemia. *Lancet* 2 (7890): 1178–1179.

Blaese, R. M., K. W. Culver, A. D. Miller, C. S. Carter, T. Fleisher, M. Clerici, G. Shearer, L. Chang, Y. Chiang, P. Tolstoshev, J. J. Greenblatt, S. A. Rosenberg, H. Klein, M. Berger, C. A. Mullen, W. J. Ramsey, L. Muul, R. A. Morgan, and W. F. Anderson. 1995. T lymphocyte-directed gene therapy for ADA- SCID: Initial trial results after 4 years. *Science* 270 (5235): 475–480.

Blagosklonny, M. V. 2007. Cancer stem cell and cancer stemloids: From biology to therapy. *Cancer Biology and Therapy* 6 (11): 1684–1690.

Blau, H. M., T. R. Brazelton, and J. M. Weimann. 2001. The evolving concept of a stem cell: Entity or function? *Cell* 105 (7): 829–841.

Bonnet, D., and J. E. Dick. 1997. Human acute myeloid leukemia is organized as a hierarchy that originates from a primitive hematopoietic cell. *Nature Medicine* 3 (7): 730–737.

Boulais, P. E., and P. S. Frenette. 2015. Making sense of hematopoietic stem cell niches. *Blood* 125 (17): 2621–2629.

Boyd, R. 1999a. Homeostasis, species, and higher taxa. In *Species: New Interdisciplinary Essays,* edited by R. Wilson. Cambridge, MA: MIT Press.

——. 1999b. Kinds, complexity and multiple realization: Comments on Millikan's "Historical Kinds and the Special Sciences." *Philosophical Studies* 95: 67–98.

Brabletz, T., A. Jung, S. Spaderna, F. Hlubek, and T. Kirchner. 2005. Opinion: Migrating cancer stem cells—an integrated concept of malignant tumour progression. *Nature Reviews Cancer* 5 (9): 744–749.

Brack, A. S., M. J. Conboy, S. Roy, M. Lee, C. J. Kuo, C. Keller, and T. A. Rando. 2007. Increased Wnt signaling during aging alters muscle stem cell fate and increases fibrosis. *Science* 317 (5839): 807–810.

Brandt, C. 2012. Stem cells, reversibility and reprogramming: historical perspectives. In *Differing Routes to Stem Cell Research: Germany and Italy,* edited by R. G. Mazzolini and H.-J. Rheinberger. Berlin: Duncker & Humblot.

Brawley, C., and E. Matunis. 2004. Regeneration of male germline stem cells by spermatogonial dedifferentiation in vivo. *Science* 304 (5675): 1331–1334.

Brazelton, T. R., F. M. Rossi, G. I. Keshet, and H. M. Blau. 2000. From marrow to brain: Expression of neuronal phenotypes in adult mice. *Science* 290 (5497): 1775–1779.

Brennan, S. K., Q. Wang, R. Tressler, C. Harley, N. Go, E. Bassett, C. A. Huff, R. J. Jones, and W. Matsui. 2010. Telomerase inhibition targets clonogenic multiple myeloma cells through telomere length-dependent and independent mechanisms. *Public Library of Science One* 5 (9): e12487.

Brinster, R. L. 1974. The effect of cells transferred into the mouse blastocyst on subsequent development. *Journal of Experimental Medicine* 140 (4): 1049–1056.

Brown, M. E., E. Rondon, D. Rajesh, A. Mack, R. Lewis, X. Feng, L. J. Zitur, R. D. Learish, and E. F. Nuwaysir. 2010. Derivation of induced pluripotent stem cells from human peripheral blood T lymphocytes. *Public Library of Science One* 5 (6): e11373.

Brown, N., A. Kraft, and P. Martin. 2006. The promissory pasts of blood stem cells. *BioSocieties* 1 (3): 329–348.

Bru, T., C. Clarke, M. J. McGrew, H. M. Sang, I. Wilmut, and J. J. Blow. 2008. Rapid induction of pluripotency genes after exposure of human somatic cells to mouse ES cell extracts. *Experimental Cell Research* 314 (14): 2634–2642.

Bruce, W. R., and H. Van Der Gaag. 1963. A quantitative assay for the number of murine lymphoma cells capable of proliferation in vivo. *Nature* 199 (4888): 79–80.

Buczacki, S., R. J. Davies, and D. J. Winton. 2011. Stem cells, quiescence and rectal carcinoma: An unexplored relationship and potential therapeutic target. *British Journal of Cancer* 105 (9): 1253–1259.

Budde, M. 1926. Ueber die Entstehung der Fetalinklusionen, komplizierten Dermoide und Teratome und ihre Beziehungen zueinander. *Gynecologic and Obstetric Investigation* 74 (5): 276–283.

Buick, R. N., M. D. Minden, and E. A. McCulloch. 1979. Self-renewal in culture of proliferative blast progenitor cells in acute myeloblastic leukemia. *Blood* 54 (1): 95–104.

Burian, R. 2005. *The Epistemology of Development, Evolution, and Genetics.* Cambridge: Cambridge University Press.

Burleigh, A. R. 2008. Of germ cells, trophoblasts, and cancer stem cells. *Integrative Cancer Therapies* 7 (4): 276–281.

Burns, C. E., and L. I. Zon. 2002. Portrait of a stem cell. *Developmental Cell* 3 (5): 612–613.

Cabarcas, S. M., L. A. Mathews, and W. L. Farrar. 2011. The cancer stem cell niche—there goes the neighborhood? *International Journal of Cancer* 129 (10): 2315–2327.

Cairns, J. 1975. Mutation selection and the natural history of cancer. *Nature* 255 (5505): 197–200.

Calabrese, C., H. Poppleton, M. Kocak, T. L. Hogg, C. Fuller, B. Hamner, E. Y. Oh, M. W. Gaber, D. Finklestein, M. Allen, A. Frank, I. T. Bayazitov, S. S. Zakharenko, A. Gajjar, A. Davidoff, and R. J. Gilbertson. 2007. A perivascular niche for brain tumor stem cells. *Cancer Cell* 11 (1): 69–82.

Camacho, L. H. 2003. Clinical applications of retinoids in cancer medicine. *Journal of Biological Regulators and Homeostatic Agents* 17 (1): 98–114.

Cambrosio, A., and P. Keating. 1992a. Between fact and technique: The beginnings of hybridoma technology. *Journal of the History of Biology* 25 (2): 175–230.

———. 1992b. A matter of FACS: Constituting novel entities in immunology. *Medical Anthropology Quarterly*, n.s., 6 (4): 362–384.

———. 1995. *Exquisite Specificity: The Monoclonal Antibody Revolution.* New York: Oxford University Press.

Campos, B., F. Wan, M. Farhadi, A. Ernst, F. Zeppernick, K. E. Tagscherer, R. Ahmadi, J. Lohr, C. Dictus, G. Gdynia, S. E. Combs, V. Goidts, B. M. Helmke, V. Eckstein, W. Roth, P. Beckhove, P. Lichter, A. Unterberg, B. Radlwimmer, and C. Herold-Mende. 2010. Differentiation therapy exerts antitumor effects on stem-like glioma cells. *Clinical Cancer Research* 16 (10): 2715–2728.

Carlson, B. M., and J. A. Faulkner. 1989. Muscle transplantation between young and old rats: Age of host determines recovery. *American Journal of Physiology* 256 (6 Pt 1): C1262–C1266.

Carmichael, H. T. 1931. Squamous epithelial rests in hypophysis cerebri. *Archives of Neurology and Psychiatry* 26: 966–975.

Cetin, I., and M. Topcul. 2012. Cancer stem cells in oncology. *Journal of Balkan Union of Oncology* 17 (4): 644–648.

Chaffer, C. L., I. Brueckmann, C. Scheel, A. J. Kaestli, P. A. Wiggins, L. O. Rodrigues, M. Brooks, F. Reinhardt, Y. Su, K. Polyak, L. M. Arendt, C. Kuperwasser, B. Bierie, and R. A. Weinberg. 2011. Normal and neoplastic nonstem cells can spontaneously convert to a stem-like state. *Proceedings of the National Academy of Sciences USA* 108 (19): 7950–7955.

Chaffer, C. L., N. D. Marjanovic, T. Lee, G. Bell, C. G. Kleer, F. Reinhardt, A. C. D'Alessio, R. A. Young, and R. A. Weinberg. 2013. Poised chromatin at the ZEB1 promoter enables breast cancer cell plasticity and enhances tumorigenicity. *Cell* 154 (1): 61–74.

Chaffer, C. L., and R. A. Weinberg. 2015. How does multistep tumorigenesis really proceed? *Cancer Discovery* 5 (1): 22–24.

Chan, K. S., I. Espinosa, M. Chao, D. Wong, L. Ailles, M. Diehn, H. Gill, J. Presti, Jr., H. Y. Chang, M. van de Rijn, L. Shortliffe, and I. L. Weissman. 2009. Identification, molecular characterization, clinical prognosis, and therapeutic targeting of human bladder tumor-initiating cells. *Proceedings of the National Academy of Sciences USA* 106 (33): 14016–14021.

Chao, M. P., A. A. Alizadeh, C. Tang, J. H. Myklebust, B. Varghese, S. Gill, M. Jan, A. C. Cha, C. K. Chan, B. T. Tan, C. Y. Park, F. Zhao, H. E. Kohrt, R. Malumbres, J. Briones, R. D. Gascoyne, I. S. Lossos, R. Levy, I. L. Weissman, and R. Majeti. 2010. Anti-CD47 antibody synergizes with rituximab to promote phagocytosis and eradicate non-Hodgkin lymphoma. *Cell* 142 (5): 699–713.

Chargari, C., C. Moncharmont, A. Levy, J. B. Guy, G. Bertrand, M. Guilbert, C. Rousseau, L. Vedrine, G. Alphonse, R. A. Toillon, C. Rodriguez-Lafrasse, E. Deutsch, and N. Magne. 2012. [Cancer stem cells, cornerstone of radioresistance and perspectives for radiosensitization: glioblastoma as an example]. *Bulletin du Cancer* 99 (12): 1153–1160.

Chen, K., Y. H. Huang, and J. L. Chen. 2013. Understanding and targeting cancer stem cells: Therapeutic implications and challenges. *Acta Pharmacologica Sinica* 34 (6): 732–740.

Chen, Z. L., W. M. Yu, and S. Strickland. 2007. Peripheral regeneration. *Annual Review of Neuroscience* 30: 209–233.

Cheng, J., N. Turkel, N. Hemati, M. T. Fuller, A. J. Hunt, and Y. M. Yamashita. 2008. Centrosome misorientation reduces stem cell division during ageing. *Nature* 456 (7222): 599–604.

Chiba, T., Y. W. Zheng, K. Kita, O. Yokosuka, H. Saisho, M. Onodera, H. Miyoshi, M. Nakano, Y. Zen, Y. Nakanuma, H. Nakauchi, A. Iwama, and H. Taniguchi. 2007. Enhanced self-renewal capability in hepatic stem/progenitor cells drives cancer initiation. *Gastroenterology* 133 (3): 937–950.

Chiquoine, A. D. 1954. The identification, origin, and migration of the primordial germ cells in the mouse embryo. *Anatomical Record* 118 (2): 135–146.

Choi, S. 2003. Improving Bird's antidotes. *Australasian Journal of Philosophy* 81 (4): 573–580.

Choi, S. M., H. Liu, P. Chaudhari, Y. Kim, L. Cheng, J. Feng, S. Sharkis, Z. Ye and Y. Y. Jang. 2011. Reprogramming of EBV-immortalized B-lymphocyte cell lines into induced pluripotent stem cells. *Blood* 118 (7): 1801–1805.

Christen, B., V. Robles, M. Raya, I. Paramonov and J. C. Izpisua Belmonte. 2010. Regeneration and reprogramming compared. *BioMed Central Biology* 8: 5.

Clarke, M. F., J. E. Dick, P. B. Dirks, C. J. Eaves, C. H. Jamieson, D. L. Jones, J. Visvader, I. L. Weissman, and G. M. Wahl. 2006. Cancer stem cells—perspectives on current status and future directions: AACR Workshop on cancer stem cells. *Cancer Research* 66 (19): 9339–9344.

Clarke, M. F., and M. Fuller. 2006. Stem cells and cancer: two faces of eve. *Cell* 124 (6): 1111–1115.

Clarkson, B., J. Fried, A. Strife, Y. Sakai, K. Ota, and T. Okita. 1970. Studies of cellular proliferation in human leukemia: 3. Behavior of leukemic cells in three adults with acute leukemia given continuous infusions of 3H-thymidine for 8 or 10 days. *Cancer* 25 (6): 1237–1260.

Clarkson, B. D., and J. Fried. 1971. Changing concepts of treatment in acute leukemia. *Medical Clinics of North America* 55 (3): 561–600.

Clein, G. P., and R. J. Flemans. 1966. Involvement of the erythroid series in blastic crisis of chronic myeloid leukaemia: Further evidence for the presence of Philadelphia chromosome in erythroblasts. *British Journal of Haematology* 12 (6): 754–758.

Clevers, H. 2006. Wnt/beta-catenin signaling in development and disease. *Cell* 127 (3): 469–480.

Clevers, H., K. M. Loh, and R. Nusse. 2014. An integral program for tissue renewal and regeneration: Wnt signaling and stem cell control. *Science* 346 (6205): 10.1126/science.1248012.

Colvin, G. A., P. J. Quesenberry, and M. S. Dooner. 2006. The stem cell continuum: A new model of stem cell regulation. *Handbook of Experimental Pharmacology* (174): 169–183.

Cohnheim, J. 1867. Ueber entzundung und eiterung. *Virchows Archiv fur Pathologische Anatomie und Physiologie und fur Klinische Medizin* 40: 1–79.

——. 1875. Congenitales, quergestreiftes Muskelsarkom der Nieren. *Virchows Archiv* 65: 64–69.

——. 1889. *General Pathology. A handbook for practitioners and students*. 2nd ed. Translation by Alexander McKee. London: The New Sydenham Society.

Collins, A. T., P. A. Berry, C. Hyde, M. J. Stower, and N. J. Maitland. 2005. Prospective identification of tumorigenic prostate cancer stem cells. *Cancer Research* 65 (23): 10946–10951.

Conboy, I. M., M. J. Conboy, G. M. Smythe, and T. A. Rando. 2003. Notch-mediated restoration of regenerative potential to aged muscle. *Science* 302 (5650): 1575–1577.

Conboy, I. M., A. J. Wagers, E. R. Girma, I. L. Weissman, and T. A. Rando. 2005. Rejuvenation of aged progenitor cells by exposure to a young systemic environment. *Nature* 433 (7027): 760–764.

Conley, S. J., E. Gheordunescu, P. Kakarala, B. Newman, H. Korkaya, A. N. Heath, S. G. Clouthier, and M. S. Wicha. 2012. Antiangiogenic agents increase breast cancer stem cells via the generation of tumor hypoxia. *Proceedings of the National Academy of Sciences USA* 109 (8): 2784–2789.

Copland, M., A. Hamilton, L. J. Elrick, J. W. Baird, E. K. Allan, N. Jordanides, M. Barow, J. C. Mountford, and T. L. Holyoake. 2006. Dasatinib (BMS-354825) targets an earlier progenitor population than imatinib in primary CML but does not eliminate the quiescent fraction. *Blood* 107 (11): 4532–4539.

Corces-Zimmerman, M. R., W. J. Hong, I. L. Weissman, B. C. Medeiros, and R. Majeti. 2014. Preleukemic mutations in human acute myeloid leukemia affect epigenetic regulators and persist in remission. *Proceedings of the National Academy of Sciences USA* 111 (7): 2548–2553.

Cortes, J., S. O'Brien, and H. Kantarjian. 2004. Discontinuation of imatinib therapy after achieving a molecular response. *Blood* 104 (7): 2204–2205.

Cozzio, A., E. Passegue, P. M. Ayton, H. Karsunky, M. L. Cleary, and I. L. Weissman. 2003. Similar MLL-associated leukemias arising from self-renewing stem cells and short-lived myeloid progenitors. *Genes and Development* 17 (24): 3029–3035.

Crist, E., and A. I. Tauber. 1999. Selfhood, immunity, and the biological imagination: the thought of Frank Macfarlane Burnet. *Biology and Philosophy* 15 (4): 509–533.

Critchley, M., and R. N. Ironside. 1926. The pituitary adamantinomata. *Brain* 49: 437–481.

Cronkite, E. P., V. P. Bond, T. M. Fliedner, and J. R. Rubini. 1959. The use of tritiated thymidine in the study of DNS synthesis and cell turnover in hemopoietic tissues. *Laboratory Investigation* 8 (1): 263–275; discussion 276–267.

Cronkite, E. P., and L. E. Feinendegen. 1976. Notions about human stem cells. *Nouvelle Revue Française d'Hématologie Blood Cells* 17 (1–2): 269–284.

Cummins, R. 1975. Functional analysis. *Journal of Philosophy* 72 (20): 741–765.

Curtis, S. J., K. W. Sinkevicius, D. Li, A. N. Lau, R. R. Roach, R. Zamponi, A. E. Woolfenden, D. G. Kirsch, K. K. Wong, and C. F. Kim. 2010. Primary tumor genotype is an important determinant in identification of lung cancer propagating cells. *Cell Stem Cell* 7 (1): 127–133.

Cushing, H. 1932. *Intracranial Tumors. Notes Upon a Series of Two Thousand Verified Cases with Surgical Mortality Percentages Pertaining Thereto.* Springfield, IL: Charles C Thomas.

Cyranoski, David. 2014a. Acid-bath stem-cell study under investigation, *Nature*, February 17: http://www.nature.com/news/acid-bath-stem-cell-study-under-investigation-1.14738.

———. 2014b. Stem-cell scientist found guilty of misconduct, *Nature*, April 1: http://www.nature.com/news/stem-cell-scientist-found-guilty-of -misconduct-1.14974.

Dalerba, P., S. J. Dylla, I. K. Park, R. Liu, X. Wang, R. W. Cho, T. Hoey, A. Gurney, E. H. Huang, D. M. Simeone, A. A. Shelton, G. Parmiani, C. Castelli, and M. F. Clarke. 2007. Phenotypic characterization of human colorectal cancer stem cells. *Proceedings of the National Academy of Sciences USA* 104 (24): 10158–10163.

Damjanov, I., and P. W. Andrews. 2007. The terminology of teratocarcinomas and teratomas. *Nature Biotechnology* 25 (11): 1212.

Darwin, C. 1875. *The Variation of Animals and Plants under Domestication*. 2nd ed. London: J. Murray.

Das, S., M. Srikanth, and J. A. Kessler. 2008. Cancer stem cells and glioma. *Nature Clinical Practice. Neurology* 4 (8): 427–35.

Dembinski, J. L., and S. Krauss. 2009. Characterization and functional analysis of a slow cycling stem cell-like subpopulation in pancreas adenocarcinoma. *Clinical and Experimental Metastasis* 26 (7): 611–623.

Deng, W., and H. Lin. 1997. Spectrosomes and fusomes anchor mitotic spindles during asymmetric germ cell divisions and facilitate the formation of a polarized microtubule array for oocyte specification in Drosophila. *Developmental Biology* 189 (1): 79–94.

Dick, J. E. 1996. Normal and leukemic human stem cells assayed in SCID mice. *Seminars in Immunology* 8 (4): 197–206.

———. 2008. Stem cell concepts renew cancer research. *Blood* 112 (13): 4793–4807.

Diehn, M., R. W. Cho, N. A. Lobo, T. Kalisky, M. J. Dorie, A. N. Kulp, D. Qian, J. S. Lam, L. E. Ailles, M. Wong, B. Joshua, M. J. Kaplan, I. Wapnir, F. M. Dirbas, G. Somlo, C. Garberoglio, B. Paz, J. Shen, S. K. Lau, S. R. Quake, J. M. Brown, I. L. Weissman, and M. F. Clarke. 2009. Association of reactive oxygen species levels and radioresistance in cancer stem cells. *Nature* 458 (7239): 780–783.

Dittmar, T., and K. S. Zänker, eds. 2008. *Cancer and Stem Cells*. New York: Nova.

———, eds. 2013. *Role of Cancer Stem Cells in Cancer Biology and Therapy*. Boca Raton, FL: CRC Press.

Dixon, F. J., and R. A. Moore. 1953. Testicular tumors: A clinicopathological study. *Cancer* 6 (3): 427–454.

Donati, G., and F. M. Watt. 2015. Stem cell heterogeneity and plasticity in epithelia. *Cell Stem Cell* 16 (5): 465–476.

Dontu, G., M. Al-Hajj, W. M. Abdallah, M. F. Clarke, and M. S. Wicha. 2003. Stem cells in normal breast development and breast cancer. *Cell Proliferation* 36 (Suppl 1): 59–72.

Drummond, M. W., N. Heaney, J. Kaeda, F. E. Nicolini, R. E. Clark, G. Wilson, P. Shepherd, J. Tighe, L. McLintock, T. Hughes, and T. L. Holyoake. 2009. A pilot study of continuous imatinib vs pulsed imatinib with or without

G-CSF in CML patients who have achieved a complete cytogenetic response. *Leukemia* 23 (6): 1199–1201.

Drummond, W. A. D. 1939. Infrasellar adamantinoma. *Proceedings of the Royal Society of Medicine* 32 (3): 200–207.

Duchesneau, F. 1987. *Genèse de la théorie cellulaire.* Paris: Vrin.

Duffy, W. C. 1920. Hypophyseal duct tumors: A report of three cases and a fourth case of cyst of rathke's pouch. *Annals of Surgery* 72: 537–555; 725–757.

Dufour, C., J. Cadusseau, P. Varlet, A. L. Surena, G. P. de Faria, A. Dias-Morais, N. Auger, N. Leonard, E. Daudigeos, C. Dantas-Barbosa, J. Grill, V. Lazar, P. Dessen, G. Vassal, V. Prevot, A. Sharif, H. Chneiweiss, and M. P. Junier. 2009. Astrocytes reverted to a neural progenitor-like state with transforming growth factor alpha are sensitized to cancerous transformation. *Stem Cells* 27 (10): 2373–2382.

Duggal, R., B. Minev, U. Geissinger, H. Wang, N. G. Chen, P. S. Koka, and A. A. Szalay. 2013. Biotherapeutic approaches to target cancer stem cells. *Journal of Stem Cells* 8 (3–4): 135–149.

Dupont, J.-C., and S. Schmitt, ed. 2004. *Du Feuillet au Gène. Une Histoire de l'Embryologie Moderne (fin XVIIIe/XXe siècle).* Paris: Edition rue d'Ulm.

Dupré, J. 1981. Natural kinds and biological taxa. *Philosophical Review* 90 (1): 66–90.

Ebbe, S. 1968. Megakaryocytopoiesis and platelet turnover. *Series Haematologica* 1 (2): 65–98.

Echeverri, K., J. D. Clarke, and E. M. Tanaka. 2001. In vivo imaging indicates muscle fiber dedifferentiation is a major contributor to the regenerating tail blastema. *Developmental Biology* 236 (1): 151–164.

Edris, B., K. Weiskopf, A. K. Volkmer, J. P. Volkmer, S. B. Willingham, H. Contreras-Trujillo, J. Liu, R. Majeti, R. B. West, J. A. Fletcher, A. H. Beck, I. L. Weissman, and M. van de Rijn. 2012. Antibody therapy targeting the CD47 protein is effective in a model of aggressive metastatic leiomyosarcoma. *Proceedings of the National Academy of Sciences USA* 109 (17): 6656–6661.

Efroni, S., R. Duttagupta, J. Cheng, H. Dehghani, D. J. Hoeppner, C. Dash, D. P. Bazett-Jones, S. Le Grice, R. D. McKay, K. H. Buetow, T. R. Gingeras, T. Misteli, and E. Meshorer. 2008. Global transcription in pluripotent embryonic stem cells. *Cell Stem Cell* 2 (5): 437–447.

Eisenberg, L. M., and C. A. Eisenberg. 2003. Stem cell plasticity, cell fusion, and transdifferentiation. *Birth Defects Research Part C: Embryo Today* 69 (3): 209–218.

Eisenhauer, E. A., P. Therasse, J. Bogaerts, L. H. Schwartz, D. Sargent, R. Ford, J. Dancey, S. Arbuck, S. Gwyther, M. Mooney, L. Rubinstein, L. Shankar, L. Dodd, R. Kaplan, D. Lacombe, and J. Verweij. 2009. New response evaluation criteria in solid tumours: Revised RECIST guideline (version 1.1). *European Journal of Cancer* 45 (2): 228–247.

Ellis, B. 2001. *Scientific Essentialism.* Cambridge: Cambridge University Press.

Eminli, S., A. Foudi, M. Stadtfeld, N. Maherali, T. Ahfeldt, G. Mostoslavsky, H. Hock, and K. Hochedlinger. 2009. Differentiation stage determines potential of hematopoietic cells for reprogramming into induced pluripotent stem cells. *Nature Genetics* 41 (9): 968–976.

Engler, A. J., S. Sen, H. L. Sweeney, and D. E. Discher. 2006. Matrix elasticity directs stem cell lineage specification. *Cell* 126 (4): 677–689.

Eramo, A., F. Lotti, G. Sette, E. Pilozzi, M. Biffoni, A. Di Virgilio, C. Conticello, L. Ruco, C. Peschle, and R. De Maria. 2008. Identification and expansion of the tumorigenic lung cancer stem cell population. *Cell Death and Differentiation* 15 (3): 504–514.

Erdheim, J. 1903. Zur normalen und pathologischen Histologie der Olantlula thyreoidea, parathyreoidea und Hypophysis. *Beiträge zur Pathologischen Anatomie und zur allgemeinen Pathologie* 33: 158–234.

——. 1904. Ueber Hypophysenganggeschwuelste und Hirncholesteatome. *Sitzber math-naturwiss Klasse Kais Akad Wissenschaft* 113 (2): 537–726.

Ereshefsky, M. 2012. Species. In *The Stanford Encyclopedia of Philosophy (Spring 2010 Edition)*, edited by E. N. Zalta: http://plato.stanford.edu/archives /spr2010/entries/species/.

Essers, M. A., S. Offner, W. E. Blanco-Bose, Z. Waibler, U. Kalinke, M. A. Duchosal, and A. Trumpp. 2009. IFNalpha activates dormant haematopoietic stem cells in vivo. *Nature* 458 (7240): 904–908.

Essers, M. A., and A. Trumpp. 2010. Targeting leukemic stem cells by breaking their dormancy. *Molecular Oncology* 4 (5): 443–450.

Evans, M. J., and M. H. Kaufman. 1981. Establishment in culture of pluripotential cells from mouse embryos. *Nature* 292 (5819): 154–156.

Evsikov, A. V., and D. Solter. 2003. Comment on "'Stemness:' Transcriptional profiling of embryonic and adult stem cells" and "a stem cell molecular signature." *Science* 302 (5644): 393.

Ewing, J. 1940. *Neoplastic Diseases.* 4th ed. Philadelphia: W. B. Saunders.

Eyler, C. E., and J. N. Rich. 2008. Survival of the fittest: cancer stem cells in therapeutic resistance and angiogenesis. *Journal of Clinical Oncology* 26 (17): 2839–45.

Fagan, M. B. 2007. The search for the hematopoietic stem cell: Social interaction and epistemic success in immunology. *Studies in History and Philosophy of Science Part C* 38 (1): 217–237.

——. 2010. Stems and standards: Social interaction in the search for blood stem cells. *Journal of the History of Biology* 43 (1): 67–109.

——. 2012. Waddington redux: Models and explanation in stem cell and systems biology. *Biology and Philosophy* 27 (2): 179–213.

——. 2013a. *Philosophy of Stem Cell Biology. Knowledge in Flesh and Blood.* London: Palgrave-Macmillan.

———. 2013b. The stem cell uncertainty principle. *Philosophy of Science* 80 (5): 945–957.

———. 2015. Crucial stem cell experiments? Stem cells, uncertainty, and single-cell experiments. *THEORIA* 30 (2): 183–205.

Farrar, W. L., ed. 2009. *Cancer Stem Cells.* New York: Cambridge University Press.

Fearon, E. R., P. J. Burke, C. A. Schiffer, B. A. Zehnbauer, and B. Vogelstein. 1986. Differentiation of leukemia cells to polymorphonuclear leuko-cytes in patients with acute nonlymphocytic leukemia. *New England Journal of Medicine* 315 (1): 15–24.

Federici, G., V. Espina, L. Liotta, and K. H. Edmiston. 2011. Breast cancer stem cells: A new target for therapy. *Oncology (Williston Park)* 25 (1): 25–28, 30.

Feitelson, M. A., A. Arzumanyan, R. J. Kulathinal, S. W. Blain, R. F. Holcombe, J. Mahajna, M. Marino, M. L. Martinez-Chantar, R. Nawroth, I. Sanchez-Garcia, D. Sharma, N. K. Saxena, N. Singh, P. J. Vlachostergios, S. Guo, K. Honoki, H. Fujii, A. G. Georgakilas, A. Amedei, E. Niccolai, A. Amin, S. S. Ashraf, C. S. Boosani, G. Guha, M. R. Ciriolo, K. Aquilano, S. Chen, S. I. Mohammed, A. S. Azmi, D. Bhakta, D. Halicka, and S. Now-sheen. 2015. Sustained proliferation in cancer: Mechanisms and novel therapeutic targets. *Seminars in Cancer Biology,* in press. doi:10.1016/j.semcancer.2015.02.006.

Fekete, E., and M. A. Ferrigno. 1952. Studies on a transplantable teratoma of the mouse. *Cancer Research* 12 (6): 438–440.

Ferrari, G., G. Cusella-De Angelis, M. Coletta, E. Paolucci, A. Stornaiuolo, G. Cossu, and F. Mavilio. 1998. Muscle regeneration by bone marrow-derived myogenic progenitors. *Science* 279 (5356): 1528–1530.

Feuring-Buske, M., B. Gerhard, J. Cashman, R. K. Humphries, C. J. Eaves, and D. E. Hogge. 2003. Improved engraftment of human acute myeloid leukemia progenitor cells in beta 2-microglobulin-deficient NOD/SCID mice and in NOD/SCID mice transgenic for human growth factors. *Leukemia* 17 (4): 760–763.

Fialkow, P. J. 1979. Clonal origin of human tumors. *Annual Review of Medicine* 30: 135–143.

———. 1980. Clonal and stem cell origin of blood cell neoplasms. In *Contemporary Hematology and Oncology,* edited by J. Lobue, A. S. Gordon, R. Silber, and F. M. Muggia. New York: Plenum.

Fialkow, P. J., S. M. Gartler, and A. Yoshida. 1967. Clonal origin of chronic myelocytic leukemia in man. *Proceedings of the National Academy of Sciences USA* 58 (4): 1468–1471.

Fialkow, P. J., R. J. Jacobson, and T. Papayannopoulou. 1977. Chronic myelo-cytic leukemia: Clonal origin in a stem cell common to the granulocyte, erythrocyte, platelet and monocyte/macrophage. *American Journal of Medicine* 63 (1): 125–130.

Fialkow, P. J., E. Klein, G. Klein, P. Clifford, and S. Singh. 1973. Immunoglobulin and glucose-6-phosphate dehydrogenase as markers of cellular origin in Burkitt lymphoma. *Journal of Experimental Medicine* 138 (1): 89–102.

Filatova, A., T. Acker, and B. K. Garvalov. 2013. The cancer stem cell niche(s): The crosstalk between glioma stem cells and their microenvironment. *Biochimica et Biophysica Acta* 1830 (2): 2496–2508.

Finch, B. W., and B. Ephrussi. 1967. Retention of multiple developmental potentialities by cells of a mouse testicular teratocarcinoma during prolonged culture in vitro and their extinction upon hybridisation with cells of permanent lines. *Proceedings of the National Academy of Sciences USA* 57 (3): 615–621.

Fischer, B. 1913. Ueber ein primaeres Adamantinom der Tibia. *Frankfurter Zeitschrift für Pathologie* 12: 422–441.

Flemming, A. 2015. Cancer stem cells: Targeting the root of cancer relapse. *Nature Reviews Drug Discovery* 14 (3): 165.

Fojo, T., and C. Grady. 2009. How much is life worth: Cetuximab, non-small cell lung cancer, and the $440 billion question. *Journal of the National Cancer Institute* 101 (15): 1044–1048.

Ford, A. M., M. B. Mansur, C. L. Furness, F. W. van Delft, J. Okamura, T. Suzuki, H. Kobayashi, Y. Kaneko, and M. Greaves. 2015. Protracted dormancy of pre-leukemic stem cells. *Leukemia* 29 (11): 2202–2207.

Ford, C. E., J. L. Hamerton, D. W. Barnes, and J. F. Loutit. 1956. Cytological identification of radiation-chimaeras. *Nature* 177 (4506): 452–454.

Fortunel, N. O., H. H. Otu, H. H. Ng, J. Chen, X. Mu, T. Chevassut, X. Li, M. Joseph, C. Bailey, J. A. Hatzfeld, A. Hatzfeld, F. Usta, V. B. Vega, P. M. Long, T. A. Libermann, and B. Lim. 2003. Comment on "'Stemness:' Transcriptional profiling of embryonic and adult stem cells" and "a stem cell molecular signature." *Science* 302 (5644): 393.

Frank, S. A. 2007. *Dynamics of Cancer: Incidence, Inheritance, and Evolution.* Princeton, NJ: Princeton University Press.

Fraser, C. C., H. Kaneshima, G. Hansteen, M. Kilpatrick, R. Hoffman, and B. P. Chen. 1995. Human allogeneic stem cell maintenance and differentiation in a long-term multilineage SCID-hu graft. *Blood* 86 (5): 1680–1693.

Frid, M. G., V. A. Kale, and K. R. Stenmark. 2002. Mature vascular endothelium can give rise to smooth muscle cells via endothelial-mesenchymal transdifferentiation: In vitro analysis. *Circulation Research* 90 (11): 1189–1196.

Gartler, S. M., L. Ziprkowski, A. Krakowski, R. Ezra, A. Szeinberg, and A. Adam. 1966. Glucose-6-phosphate dehydrogenase mosaicism as a tracer in the study of hereditary multiple trichoepithelioma. *American Journal of Human Genetics* 18 (3): 282–287.

Gavosto, F., G. Maraini, and A. Pileri. 1960. Proliferative capacity of acute leukaemia cells. *Nature* 187: 611–612.

Gavosto, F., A. Pileri, C. Bachi, and L. Pegoraro. 1964. Proliferation and maturation defect in acute leukaemia cells. *Nature* 203: 92–94.

Gavosto, F., A. Pileri, V. Gabutti, and P. Masera. 1967a. Non-self-maintaining kinetics of proliferating blasts in human acute leukaemia. *Nature* 216 (5111): 188–189.

———. 1967b. Cell population kinetics in human acute leukaemia. *European Journal of Cancer* 3 (4): 301–307.

Gazave, E., J. Béhague, L. Laplane, A. Guillou, L. Préau, A. Demilly, G. Balavoine, and M. Vervoort. 2013. Posterior elongation in the annelid Platynereis dumerilii involves stem cells molecularly related to primordial germ cells. *Developmental Biology* 382 (1): 246–267.

Gerby, B. 2010. Identification des cellules initiant les LAL-T humaines dans les souris immunodéficientes et étude de la modulation de l'expression de l'oncogène SCL/TAL1 sur le développement leucémique. PhD diss., University Paris 7, Paris.

Germain, P.-L. 2012. Cancer cells and adaptive explanations. *Biology and Philosophy* 27 (6): 785–810.

Ghiaur, G., J. Gerber, and R. J. Jones. 2012. Concise review: Cancer stem cells and minimal residual disease. *Stem Cells* 30 (1): 89–93.

Gibbs, C. P., V. G. Kukekov, J. D. Reith, O. Tchigrinova, O. N. Suslov, E. W. Scott, S. C. Ghivizzani, T. N. Ignatova, and D. A. Steindler. 2005. Stem-like cells in bone sarcomas: Implications for tumorigenesis. *Neoplasia* 7 (11): 967–976.

Gisselsson, D. 2007. Cancer stem cells: differentiation block or developmental back-tracking? *Seminars in Cancer Biology* 17 (3): 189–190.

Godfrey-Smith, P. 2009. *Darwinian Populations and Natural Selection.* New York: Oxford University Press.

Golan-Mashiach, M., J. E. Dazard, S. Gerecht-Nir, N. Amariglio, T. Fisher, J. Jacob-Hirsch, B. Bielorai, S. Osenberg, O. Barad, G. Getz, A. Toren, G. Rechavi, J. Itskovitz-Eldor, E. Domany, and D. Givol. 2005. Design principle of gene expression used by human stem cells: implication for pluripotency. *Federation of American Societies for Experimental Biology Journal* 19 (1): 147–149.

Goldman, J. 2005. Monitoring minimal residual disease in BCR-ABL-positive chronic myeloid leukemia in the imatinib era. *Current Opinion in Hematology* 12 (1): 33–39.

Goodell, M. A. 2003. Stem-cell "plasticity": Befuddled by the muddle. *Current Opinion in Hematology* 10 (3): 208–213.

Grácio, F., J. Cabral, and B. Tidor. 2013. Modeling stem cell induction processes. *Public Library of Science One* 8 (5): e60240.

Grafi, G. 2004. How cells dedifferentiate: A lesson from plants. *Developmental Biology* 268 (1): 1–6.

Graham, S. M., H. G. Jorgensen, E. Allan, C. Pearson, M. J. Alcorn, L. Richmond, and T. L. Holyoake. 2002. Primitive, quiescent, Philadelphia-positive stem cells from patients with chronic myeloid leukemia are insensitive to STI571 in vitro. *Blood* 99 (1): 319–325.

Greaves, M. 2000. *Cancer: The Evolutionary Legacy.* Oxford: Oxford University Press.

———. 2010. Cancer stem cells: Back to Darwin? *Seminars in Cancer Biology* 20 (2): 65–70.

Griffin, J. D., and B. Löwenberg. 1986. Clonogenic cells in acute myeloblastic leukemia. *Blood* 68 (6): 1185–1195.

Griffiths, P. 1999. Squaring the circle: natural kinds with historical essences. In *Species: New Interdisciplinary Studies,* edited by R. Wilson. Cambridge, MA: MIT Press.

Grogg, M. W., M. K. Call, M. Okamoto, M. N. Vergara, K. Del Rio-Tsonis, and P. A. Tsonis. 2005. BMP inhibition-driven regulation of six-3 underlies induction of newt lens regeneration. *Nature* 438 (7069): 858–862.

Gul-Uludag, H., J. Valencia-Serna, C. Kucharski, L. A. Marquez-Curtis, X. Jiang, L. Larratt, A. Janowska-Wieczorek, and H. Uludag. 2014. Polymeric nanoparticle-mediated silencing of CD44 receptor in CD34 acute myeloid leukemia cells. *Leukemia Research* 38 (11): 1299–1308.

Gupta, P. B., C. M. Fillmore, G. Jiang, S. D. Shapira, K. Tao, C. Kuperwasser, and E. S. Lander. 2011. Stochastic state transitions give rise to phenotypic equilibrium in populations of cancer cells. *Cell* 146 (4): 633–644.

Gupta, P. B., T. T. Onder, G. Jiang, K. Tao, C. Kuperwasser, R. A. Weinberg, and E. S. Lander. 2009. Identification of selective inhibitors of cancer stem cells by high-throughput screening. *Cell* 138 (4): 645–659.

Gurchot, C. 1975. The trophoblast theory of cancer (John Beard, 1857–1924) revisited. *Oncology* 31 (5–6): 310–333.

Haber, D. A., N. S. Gray, and J. Baselga. 2011. The evolving war on cancer. *Cell* 145 (1): 19–24.

Hambardzumyan, D., O. J. Becher, M. K. Rosenblum, P. P. Pandolfi, K. Manova-Todorova, and E. C. Holland. 2008. PI3K pathway regulates survival of cancer stem cells residing in the perivascular niche following radiation in medulloblastoma in vivo. *Genes and Development* 22 (4): 436–448.

Hamburger, A. W., and S. E. Salmon. 1977. Primary bioassay of human tumor stem cells. *Science* 197 (4302): 461–463.

Hanahan, D., and R. A. Weinberg. 2000. The hallmarks of cancer. *Cell* 100 (1): 57–70.

———. 2011. Hallmarks of cancer: the next generation. *Cell* 144 (5): 646–674.

Hanna, J., S. Markoulaki, P. Schorderet, B. W. Carey, C. Beard, M. Wernig, M. P. Creyghton, E. J. Steine, J. P. Cassady, R. Foreman, C. J. Lengner, J. A. Dausman, and R. Jaenisch. 2008. Direct reprogramming of terminally differentiated mature B lymphocytes to pluripotency. *Cell* 133 (2): 250–264.

Hanna, J., K. Saha, B. Pando, J. van Zon, C. J. Lengner, M. P. Creyghton, A. van Oudenaarden, and R. Jaenisch. 2009. Direct cell reprogramming is a stochastic process amenable to acceleration. *Nature* 462 (7273): 595–601.

Hannover, A. 1852. *Das Epithelioma, eine eigenthuemliche Geschwulst, die man im allgemeinen bisher als Krebs angesehen hat.* Leipzig: Leopold Voss.

Hauskeller, C. 2005. Science in touch: Functions of biomedical terminology. *Biology and Philosophy* 20: 815–835.

Hawley, R. G., and D. A. Sobieski. 2002. Somatic stem cell plasticity: To be or not to be. *Stem Cells* 20 (3): 195–197.

Hayat, M. A., series ed. 2012–2014. *Stem Cells and Cancer Stem Cells: Therapeutic Applications in Disease and Injury.* 12 vols. New York: Springer.

Hazelbag, H. M., G. J. Fleuren, L. J. vd Broek, A. H. Taminiau, and P. C. Hogendoorn. 1993. Adamantinoma of the long bones: Keratin subclass immunoreactivity pattern with reference to its histogenesis. *American Journal of Surgical Pathology* 17 (12): 1225–1233.

Heissig, B., K. Hattori, S. Dias, M. Friedrich, B. Ferris, N. R. Hackett, R. G. Crystal, P. Besmer, D. Lyden, M. A. Moore, Z. Werb, and S. Rafii. 2002. Recruitment of stem and progenitor cells from the bone marrow niche requires MMP-9 mediated release of kit-ligand. *Cell* 109 (5): 625–637.

Hemmati, H. D., I. Nakano, J. A. Lazareff, M. Masterman-Smith, D. H. Geschwind, M. Bronner-Fraser, and H. I. Kornblum. 2003. Cancerous stem cells can arise from pediatric brain tumors. *Proceedings of the National Academy of Sciences USA* 100 (25): 15178–15183.

Heppner, G. H., and B. E. Miller. 1983. Tumor heterogeneity: Biological implications and therapeutic consequences. *Cancer and Metastasis Reviews* 2 (1): 5–23.

Hill, R. P., and R. Perris. 2007. "Destemming" cancer stem cells. *Journal of the National Cancer Institute* 99 (19): 1435–1440.

Hirschmann-Jax, C., A. E. Foster, G. G. Wulf, J. G. Nuchtern, T. W. Jax, U. Gobel, M. A. Goodell, and M. K. Brenner. 2004. A distinct "side population" of cells with high drug efflux capacity in human tumor cells. *Proceedings of the National Academy of Sciences USA* 101 (39): 14228–14233.

Ho, H. J., T. I. Lin, H. H. Chang, S. B. Haase, S. Huang, and S. Pyne. 2012. Parametric modeling of cellular state transitions as measured with flow cytometry. *BioMed Central Bioinformatics* 13 (Suppl 5): S5.

Hochedlinger, K., and R. Jaenisch. 2002. Monoclonal mice generated by nuclear transfer from mature B and T donor cells. *Nature* 415 (6875): 1035–1038.

Hoffmann-Kolss, V. 2010. *The Metaphysics of Extrinsic Properties.* Heusenstamm: Ontos Verlag.

Holtz, M. S., S. J. Forman, and R. Bhatia. 2005. Nonproliferating CML CD34+ progenitors are resistant to apoptosis induced by a wide range of proapoptotic stimuli. *Leukemia* 19 (6): 1034–1041.

Hong, D., R. Gupta, P. Ancliff, A. Atzberger, S. Brown, S. Soneji, J. Green, S. Colman, W. Piacibello, V. Buckle, S. Tsuzuki, M. Greaves, and T. Enver. 2008. Initiating and cancer-propagating cells in TEL-AML1-associated childhood leukemia. *Science* 319 (5861): 336–9.

Hoquet, T. 2009. *Darwin contre Darwin*. Paris: Seuil.

———. 2010. Why terms matter to biological theories: The term origin as used by Darwin. *Bionomina* 1: 58–60.

Huang, M. M., and J. Zhu. 2012. The regulation of normal and leukemic hematopoietic stem cells by niches. *Cancer Microenvironment* 5 (3): 295–305.

Huang, S. 2010. Cell lineage determination in state space: A systems view brings flexibility to dogmatic canonical rules. *Public Library of Science Biology* 8 (5): e1000380.

———. 2011. Systems biology of stem cells: Three useful perspectives to help overcome the paradigm of linear pathways. *Philosophical Transactions of the Royal Society of London. Series B, Biological Sciences* 366 (1575): 2247–2259.

———. 2013. Genetic and non-genetic instability in tumor progression: link between the fitness landscape and the epigenetic landscape of cancer cells. *Cancer and Metastasis Reviews* 32 (3–4): 423–448.

Hull, D. 1965. The effect of essentialism on taxonomy: Two thousand years of stasis (1). *British Journal for the Philosophy of Science* 15 (60): 314–326.

———. 1978. A matter of individuality. *Philosophy of Science* 45: 335–360.

Humberstone, I. L. 1996. Intrinsic/extrinsic. *Synthese* 108 (2): 205–267.

Huntly, B. J., and D. G. Gilliland. 2005. Leukaemia stem cells and the evolution of cancer-stem-cell research. *Nature Reviews Cancer* 5 (4): 311–321.

Huntly, B. J., H. Shigematsu, K. Deguchi, B. H. Lee, S. Mizuno, N. Duclos, R. Rowan, S. Amaral, D. Curley, I. R. Williams, K. Akashi, and D. G. Gilliland. 2004. MOZ-TIF2, but not BCR-ABL, confers properties of leukemic stem cells to committed murine hematopoietic progenitors. *Cancer Cell* 6 (6): 587–596.

Huss, W. J., D. R. Gray, N. M. Greenberg, J. L. Mohler, and G. J. Smith. 2005. Breast cancer resistance protein-mediated efflux of androgen in putative benign and malignant prostate stem cells. *Cancer Research* 65 (15): 6640–6650.

Inoue, K., H. Wakao, N. Ogonuki, H. Miki, K.-i. Seino, R. Nambu-Wakao, S. Noda, H. Miyoshi, H. Koseki, M. Taniguchi, and A. Ogura. 2005. Generation of cloned mice by direct nuclear transfer from natural killer T Cells. *Current Biology* 15 (12): 1114–1118.

Ishida, T., T. Iijima, F. Kikuchi, T. Kitagawa, T. Tanida, T. Imamura, and R. Machinami. 1992. A clinicopathological and immunohistochemical study of osteofibrous dysplasia, differentiated adamantinoma, and adamantinoma of long bones. *Skeletal Radiology* 21 (8): 493–502.

Ito, K., R. Bernardi, A. Morotti, S. Matsuoka, G. Saglio, Y. Ikeda, J. Rosenblatt, D. E. Avigan, J. Teruya-Feldstein, and P. P. Pandolfi. 2008. PML targeting eradicates quiescent leukaemia-initiating cells. *Nature* 453 (7198): 1072–1078.

Ito, K., R. Bernardi, and P. P. Pandolfi. 2009. A novel signaling network as a critical rheostat for the biology and maintenance of the normal stem cell and the cancer-initiating cell. *Current Opinion in Genetics & Development* 19 (1): 51–59.

Ivanova, N. B., J. T. Dimos, C. Schaniel, J. A. Hackney, K. A. Moore, and I. R. Lemischka. 2002. A stem cell molecular signature. *Science* 298 (5593): 601–604.

Ivanova, N. B., J. T. Dimos, C. Schaniel, J. A. Hackney, K. A. Moore, M. Ramalho-Santos, S. Yoon, Y. Matsuzaki, R. C. Mulligan, D. A. Melton, and I. R. Lemischka. 2003. Response to comments on "'Stemness': Transcriptional profiling of embryonic and adult stem cells" and "A stem cell molecular signature." *Science* 302 (5644): 393d.

Jackson, E. B., and A. M. Brues. 1941. Studies on transplantable embryoma of mouse. *Cancer Research* 1: 494–498.

Jackson, K. A., T. Mi, and M. A. Goodell. 1999. Hematopoietic potential of stem cells isolated from murine skeletal muscle. *Proceedings of the National Academy of Sciences USA* 96 (25): 14482–14486.

Jacob, F. 1977. Mouse teratocarcinoma and embryonic antigens. *Immunological Reviews* 33: 3–32.

Jain, D., V. K. Jain, R. K. Vasishta, P. Ranjan, and Y. Kumar. 2008. Adamantinoma: A clinicopathological review and update. *Diagnostic Pathology* 3: 8.

Jaiswal, S., M. P. Chao, R. Majeti, and I. L. Weissman. 2010. Macrophages as mediators of tumor immunosurveillance. *Trends in Immunology* 31 (6): 212–219.

Jaiswal, S., C. H. Jamieson, W. W. Pang, C. Y. Park, M. P. Chao, R. Majeti, D. Traver, N. van Rooijen, and I. L. Weissman. 2009. CD47 is upregulated on circulating hematopoietic stem cells and leukemia cells to avoid phagocytosis. *Cell* 138 (2): 271–285.

Jarriault, S., Y. Schwab, and I. Greenwald. 2008. A Caenorhabditis elegans model for epithelial-neuronal transdifferentiation. *Proceedings of the National Academy of Sciences USA* 105 (10): 3790–3795.

Jiang, Y., B. N. Jahagirdar, R. L. Reinhardt, R. E. Schwartz, C. D. Keene, X. R. Ortiz-Gonzalez, M. Reyes, T. Lenvik, T. Lund, M. Blackstad, J. Du, S. Aldrich, A. Lisberg, W. C. Low, D. A. Largaespada, and C. M. Verfaillie. 2002. Pluripotency of mesenchymal stem cells derived from adult marrow. *Nature* 418 (6893): 41–49.

Jopling, C., S. Boue, and J. C. Izpisua Belmonte. 2011. Dedifferentiation, transdifferentiation and reprogramming: Three routes to regeneration. *Nature Reviews Molecular Cell Biology* 12 (2): 79–89.

Jopling, C., E. Sleep, M. Raya, M. Marti, A. Raya, and J. C. Izpisua Belmonte. 2010. Zebrafish heart regeneration occurs by cardiomyocyte dedifferentiation and proliferation. *Nature* 464 (7288): 606–609.

Jordan, M. E., ed. 2010. *Cancer Stem Cells. Cancer Etiology, Diagnosis and Treatments.* New York: Nova Science Publishers.

Joseph, I., R. Tressler, E. Bassett, C. Harley, C. M. Buseman, P. Pattamatta, W. E. Wright, J. W. Shay, and N. F. Go. 2010. The telomerase inhibitor imetelstat depletes cancer stem cells in breast and pancreatic cancer cell lines. *Cancer Research* 70 (22): 9494–9504.

Jundt, G., K. Remberger, A. Roessner, A. Schulz and K. Bohndorf. 1995. Adamantinoma of long bones: A histopathological and immunohistochemical study of 23 cases. *Pathology—Research and Practice* 191 (2): 112–120.

Kai, T., and A. Spradling. 2004. Differentiating germ cells can revert into functional stem cells in Drosophila melanogaster ovaries. *Nature* 428 (6982): 564–569.

Kamel-Reid, S., M. Letarte, C. Sirard, M. Doedens, T. Grunberger, G. Fulop, M. H. Freedman, R. A. Phillips, and J. E. Dick. 1989. A model of human acute lymphoblastic leukemia in immune-deficient SCID mice. *Science* 246 (4937): 1597–1600.

Kapse-Mistry, S., T. Govender, R. Srivastava, and M. Yergeri. 2014. Nanodrug delivery in reversing multidrug resistance in cancer cells. *Frontiers in Pharmacology* 5: 159.

Kasai, T., L. Chen, A. Mizutani, T. Kudoh, H. Murakami, L. Fu, and M. Seno. 2014. Cancer stem cells converted from pluripotent stem cells and the cancerous niche. *Journal of Stem Cells and Regenerative Medicine* 10 (1): 2–7.

Keating, P., and A. Cambrosio. 1994. "Ours is an engineering approach": Flow cytometry and the constitution of human T-cell subsets. *Journal of the History of Biology* 27: 449–479.

Keinanen, M., J. D. Griffin, C. D. Bloomfield, J. Machnicki, and A. de la Chapelle. 1988. Clonal chromosomal abnormalities showing multiple-cell-lineage involvement in acute myeloid leukemia. *New England Journal of Medicine* 318 (18): 1153–1158.

Kelly, P. N., A. Dakic, J. M. Adams, S. L. Nutt, and A. Strasser. 2007. Tumor growth need not be driven by rare cancer stem cells. *Science* 317 (5836): 337.

Kennedy, J. A., F. Barabe, A. G. Poeppl, J. C. Wang, and J. E. Dick. 2007. Comment on "Tumor growth need not be driven by rare cancer stem cells." *Science* 318 (5857): 1722.

Khamara, E. J. 2006. *Space, Time, and Theology in the Leibniz-Newton Controversy.* Heusenstamm: Ontos Verlag.

Kiger, A. A., D. L. Jones, C. Schulz, M. B. Rogers, and M. T. Fuller. 2001. Stem cell self-renewal specified by JAK-STAT activation in response to a support cell cue. *Science* 294 (5551): 2542–2545.

Kikuchi, K., J. E. Holdway, A. A. Werdich, R. M. Anderson, Y. Fang, G. F. Egnaczyk, T. Evans, C. A. Macrae, D. Y. Stainier, and K. D. Poss. 2010. Primary contribution to zebrafish heart regeneration by gata4(+) cardiomyocytes. *Nature* 464 (7288): 601–605.

Killmann, S. A. 1965. Proliferative activity of blast cells in leukemia and myelofibrosis: Morphological differences between proliferating and non-proliferating blast cells. *Acta Medica Scandinavica* 178 (3): 263–280.

——. 1968. Acute leukaemia: Development, remission/relapse pattern, relationship between normal and leukaemic haemopoiesis, and the "sleeper-to-feeder" stem cell hypothesis. *Series Haematologica* 1: 103–128.

Killmann, S. A., E. P. Cronkite, T. M. Fliedner, and V. P. Bond. 1962. Cell proliferation in multiple myeloma studied with tritiated thymidine in vivo. *Laboratory Investigation* 11: 845–853.

Kim, D., J. Wang, S. B. Willingham, R. Martin, G. Wernig, and I. L. Weissman. 2012. Anti-CD47 antibodies promote phagocytosis and inhibit the growth of human myeloma cells. *Leukemia* 26 (12): 2538–2545.

Kim, J. 1973. Causation, nomic subsumption, and the concept of event. *Journal of Philosophy* 70 (8): 217–236.

Kistler, M. 2002. L'identité des propriétés et la nécessité des lois de la nature. *Cahier de Philosophie de l'Université de Caen* 38/39: 249–273.

Kleinsmith, L. J., and G. B. Pierce. 1964. Multipotentiality of single embryonal carcinoma cells. *Cancer Research* 24: 1544–1551.

Knapp, D., and E. M. Tanaka. 2012. Regeneration and reprogramming. *Current Opinion in Genetics and Development* 22 (5): 485–493.

Koelliker, A. 1868. *Éléments d'histologie humaine.* 5ème édition. Traduction par Marc Sée. Paris: Victor Masson et fils.

——. 1882. *Embryologie ou traité complet du développement de l'homme et des animaux supérieurs.* 2ème édition. Paris: C. Reinwald.

Köhler, G., and C. Milstein. 1975. Continuous cultures of fused cells secreting antibody of predefined specificity. *Nature* 256 (5517): 495–497.

Kolata, Gina. 2009. Advances elusive in the drive to cure cancer. *New York Times,* April 24: http://www.nytimes.com/2009/04/24/health/policy/24cancer.html?pagewanted=all&_r=0.

Komarova, N. 2004. Does cancer solve an optimization problem? *Cell Cycle* 3 (7): 840–844.

Komarova, N. L., and D. Wodarz. 2004. The optimal rate of chromosome loss for the inactivation of tumor suppressor genes in cancer. *Proceedings of the National Academy of Sciences USA* 101 (18): 7017–7021.

Kondo, M., I. L. Weissman, and K. Akashi. 1997. Identification of clonogenic common lymphoid progenitors in mouse bone marrow. *Cell* 91 (5): 661–672.

Kraft, A. 2009. Manhattan transfer: Lethal radiation, bone marrow transplantation, and the birth of stem cell biology, ca. 1942–1961. *Historical Studies in the Natural Sciences* 39 (2): 171–218.

——. 2011. Converging histories, reconsidered potentialities: The stem cell and cancer. *BioSocieties* 6: 195–216.

Kreso, A., and J. E. Dick. 2014. Evolution of the cancer stem cell model. *Cell Stem Cell* 14 (3): 275–291.

Kreso, A., C. A. O'Brien, P. van Galen, O. I. Gan, F. Notta, A. M. Brown, K. Ng, J. Ma, E. Wienholds, C. Dunant, A. Pollett, S. Gallinger, J. McPherson, C. G. Mullighan, D. Shibata and J. E. Dick. 2013. Variable clonal repopulation dynamics influence chemotherapy response in colorectal cancer. *Science* 339 (6119): 543–548.

Krieger, T., and B. D. Simons. 2015. Dynamic stem cell heterogeneity. *Development* 142 (8): 1396–1406.

Krivtsov, A. V., D. Twomey, Z. Feng, M. C. Stubbs, Y. Wang, J. Faber, J. E. Levine, J. Wang, W. C. Hahn, D. G. Gilliland, T. R. Golub, and S. A. Armstrong. 2006. Transformation from committed progenitor to leukaemia stem cell initiated by MLL-AF9. *Nature* 442 (7104): 818–822.

Kuang, S., K. Kuroda, F. Le Grand, and M. A. Rudnicki. 2007. Asymmetric self-renewal and commitment of satellite stem cells in muscle. *Cell* 129 (5): 999–1010.

Kubagawa, H., L. B. Vogler, J. D. Capra, M. E. Conrad, A. R. Lawton, and M. D. Cooper. 1979. Studies on the clonal origin of multiple myeloma: Use of individually specific (idiotype) antibodies to trace the oncogenic event to its earliest point of expression in B-cell differentiation. *Journal of Experimental Medicine* 150 (4): 792–807.

Kuroda, Y., M. Kitada, S. Wakao, K. Nishikawa, Y. Tanimura, H. Makinoshima, M. Goda, H. Akashi, A. Inutsuka, A. Niwa, T. Shigemoto, Y. Nabeshima, T. Nakahata, Y.-I. Nabeshima, Y. Fujiyoshi, and M. Dezawa. 2010. Unique multipotent cells in adult human mesenchymal cell populations. *Proceedings of the National Academy of Sciences USA* 107 (19): 8639–8643.

LaBarge, M. A., and H. M. Blau. 2002. Biological progression from adult bone marrow to mononucleate muscle stem cell to multinucleate muscle fiber in response to injury. *Cell* 111 (4): 589–601.

Lagasse, E., H. Connors, M. Al-Dhalimy, M. Reitsma, M. Dohse, L. Osborne, X. Wang, M. Finegold, I. L. Weissman, and M. Grompe. 2000. Purified hematopoietic stem cells can differentiate into hepatocytes in vivo. *Nature Medicine* 6 (11): 1229–1234.

Lajtha, L. G. 1966. Cytokinetics and regulation of progenitor cells. *Journal of Cellular Physiology* 67 (3, Suppl 1): 133–148.

Lander, A. D. 2009. The "stem cell" concept: Is it holding us back? *Journal of Biology* 8 (8): 70.

Landsberg, J., J. Kohlmeyer, M. Renn, T. Bald, M. Rogava, M. Cron, M. Fatho, V. Lennerz, T. Wolfel, M. Holzel, and T. Tuting. 2012. Melanomas resist T-cell therapy through inflammation-induced reversible dedifferentiation. *Nature* 490 (7420): 412–416.

Lane, S. W., Y. J. Wang, C. Lo Celso, C. Ragu, L. Bullinger, S. M. Sykes, F. Ferraro, S. Shterental, C. P. Lin, D. G. Gilliland, D. T. Scadden, S. A. Armstrong, and D. A. Williams. 2011. Differential niche and Wnt requirements during acute myeloid leukemia progression. *Blood* 118 (10): 2849–2856.

Lansdorp, P. M. 1997. Self-renewal of stem cells. *Biology of Blood and Marrow Transplantation* 3 (4): 171–178.

Lapidot, T., C. Sirard, J. Vormoor, B. Murdoch, T. Hoang, J. Caceres-Cortes, M. Minden, B. Paterson, M. A. Caligiuri, and J. E. Dick. 1994. A cell initiating human acute myeloid leukaemia after transplantation into SCID mice. *Nature* 367 (6464): 645–648.

Laplane, L. 2011. Stem cells and the temporal boundaries of development: Toward a species-dependent view. *Biological Theory* 6 (1): 48–58.

———. 2014. Identifying some theories in developmental biology: The case of the cancer stem cell theory. In *Toward a Theory of Development*, edited by A. Minelli, and T. Pradeu. Oxford: Oxford University Press.

———. 2015. Stem cell epistemological issues. In *Stem Cell Biology and Regenerative Medicine*, edited by P. Charbord and C. Durand. Aalborg: River Publishers.

Laplane, L., A. Beke, W. Vainchenker, and E. Solary. 2015. Induced pluripotent stem cells as new model systems in oncology. *Stem Cells*, in press. doi: 10.1002/stem.2099.

Lapter, S., I. Livnat, A. Faerman, and D. Zipori. 2007. Structure and implied functions of truncated B-cell receptor mRNAs in early embryo and adult mesenchymal stem cells: Cdelta replaces Cmu in mu heavy chain-deficient mice. *Stem Cells* 25 (3): 761–770.

Larochelle, A., J. Vormoor, H. Hanenberg, J. C. Wang, M. Bhatia, T. Lapidot, T. Moritz, B. Murdoch, X. L. Xiao, I. Kato, D. A. Williams, and J. E. Dick. 1996. Identification of primitive human hematopoietic cells capable of repopulating NOD/SCID mouse bone marrow: Implications for gene therapy. *Nature Medicine* 2 (12): 1329–1337.

Larochelle, A., J. Vormoor, T. Lapidot, G. Sher, T. Furukawa, Q. Li, L. D. Shultz, N. F. Olivieri, G. Stamatoyannopoulos, and J. E. Dick. 1995. Engraftment of immune-deficient mice with primitive hematopoietic cells from beta-thalassemia and sickle cell anemia patients: Implications for evaluating human gene therapy protocols. *Human Molecular Genetics* 4 (2): 163–172.

Lawrence, H. J. 2004. Stem cells and the Heisenberg uncertainty principle. *Blood* 104 (3): 597–98.

Lawson, D. A., and O. N. Witte. 2007. Stem cells in prostate cancer initiation and progression. *Journal of Clinical Investigation* 117 (8): 2044–2050.

Lechler, T., and E. Fuchs. 2005. Asymmetric cell divisions promote stratification and differentiation of mammalian skin. *Nature* 437 (7056): 275–280.

Lee, J. T., and M. Herlyn. 2007. Old disease, new culprit: tumor stem cells in cancer. *Journal of Cellular Physiology* 213 (3): 603–609.

Legros, C., and E. Magigot. 1873. *Origine et formation du follicule dentaire chez les mammifères ("Contributions à l'étude du développement des dents", première mémoire)*. Paris: Germer Baillière.

———. 1879. *Morphologie du follicule dentaire chez les mammifères*. Paris: Germer Baillière.

Lemischka, I. 2002. Rethinking somatic stem cell plasticity. *Nature Biotechnology* 20 (5): 425.

Lengauer, C., K. W. Kinzler, and B. Vogelstein. 1998. Genetic instabilities in human cancers. *Nature* 396 (6712): 643–649.

Lenz, H. J. 2008. Colon cancer stem cells: A new target in the war against cancer. *Gastrointestinal Cancer Research* 2 (4): 203–204.

Lewis, D. 1983. Extrinsic properties. *Philosophical Studies* 44 (2): 197–200.

———. 1986. *On the Plurality of Worlds*. Oxford: Basil Blackwell.

———. 1997. Finkish Dispositions. *Philosophical Quarterly* 47 (187): 143–158.

Lewis, D. D. 1910. A contribution to the subject of tumors of the hypophysis. *Journal of the American Medical Association* 55 (12): 1002–1008.

Lewis, R. A. 2001. *Discovery: Windows on the Life Sciences*. Malden: Blackwell.

Leychkis, Y., S. R. Munzer, and J. L. Richardson. 2009. What is stemness? *Studies in History and Philosophy of Science Part C: Studies in History and Philosophy of Biological and Biomedical Sciences* 40 (4): 312–320.

Li, C., D. G. Heidt, P. Dalerba, C. F. Burant, L. Zhang, V. Adsay, M. Wicha, M. F. Clarke, and D. M. Simeone. 2007. Identification of pancreatic cancer stem cells. *Cancer Research* 67 (3): 1030–1037.

Liang, L., and J. R. Bickenbach. 2002. Somatic epidermal stem cells can produce multiple cell lineages during development. *Stem Cells* 20 (1): 21–31.

Lichtenstein, L. 1977. *Bone Tumors*. 5th ed. Saint Louis, MO: Mosby.

Lichtman, M. A. 2001. The stem cell in the pathogenesis and treatment of myelogenous leukemia: A perspective. *Leukemia* 15 (10): 1489–1494.

Linder, D., and S. M. Gartler. 1965. Glucose-6-phosphate dehydrogenase mosaicism: Utilization as a cell marker in the study of leiomyomas. *Science* 150 (692): 67–69.

———. 1967. Problem of single cell versus multicell origin of a tumor. *Proceedings of the Fifth Berkeley Symposium on Mathematical Statistics and Probability* 4: 625–633.

Liu, G., X. Yuan, Z. Zeng, P. Tunici, H. Ng, I. R. Abdulkadir, L. Lu, D. Irvin, K. L. Black, and J. S. Yu. 2006. Analysis of gene expression and chemoresistance of CD133+ cancer stem cells in glioblastoma. *Molecular Cancer* 5: 67.

Liu, L. 2001. Cloning efficiency and differentiation. *Nature Biotechnology* 19 (5): 406–406.

Livet, J., T. A. Weissman, H. Kang, R. W. Draft, J. Lu, R. A. Bennis, J. R. Sanes, and J. W. Lichtman. 2007. Transgenic strategies for combinatorial expression of fluorescent proteins in the nervous system. *Nature* 450 (7166): 56–62.

Loeb, K. R., and L. A. Loeb. 2000. Significance of multiple mutations in cancer. *Carcinogenesis* 21 (3): 379–385.

Loeb, L. A. 1991. Mutator phenotype may be required for multistage carcinogenesis. *Cancer Research* 51 (12): 3075–3079.

Loeffler, M., and I. Roeder. 2002. Tissue stem cells: Definition, plasticity, heterogeneity, self-organization and models—A conceptual approach. *Cells Tissues Organs* 171 (1): 8–26.

Lohmann, J. U. 2008. Plant stem cells: Divide et impera. In *Stem Cells: From Hydra to Man,* edited by T. C. Bosch. Dordrecht: Springer.

Lorenz, E., D. Uphoff, T. R. Reid, and E. Shelton. 1951. Modification of irradiation injury in mice and guinea pigs by bone marrow injections. *Journal of the National Cancer Institute* 12 (1): 197–201.

Löwy, I. 1991. The immunological construction of the self. In *Organism and the Origins of Self,* edited by A. I. Tauber. Dordrecht: Springer.

Madrigal, Alexis. 2008. Cancer stem cells could cause tumors, be key to cure. *Wired,* June 16: http://www.wired.com/2008/06/cancer-stem-cel/.

Maehle, A. H. 2011. Ambiguous cells: The emergence of the stem cell concept in the nineteenth and twentieth centuries. *Notes and Records of the Royal Society of London* 65 (4): 359–378.

Maenhaut, C., J. E. Dumont, P. P. Roger, and W. C. van Staveren. 2010. Cancer stem cells: A reality, a myth, a fuzzy concept or a misnomer? An analysis. *Carcinogenesis* 31 (2): 149–158.

Magee, J. A., E. Piskounova, and S. J. Morrison. 2012. Cancer stem cells: Impact, heterogeneity, and uncertainty. *Cancer Cell* 21 (3): 283–296.

Magitot, E. 1860. *Mémoire sur les tumeurs du périoste dentaire.* Paris: J-B Baillière.

Maienschein, J. 2003. *Whose View of Life? Embryos, Cloning and Stem Cells.* Cambridge, MA: Harvard University Press.

———. 2014. *Embryos under the Microscope: The Diverging Meanings of Life.* Cambridge, MA: Harvard University Press.

Majeti, R., M. P. Chao, A. A. Alizadeh, W. W. Pang, S. Jaiswal, K. D. Gibbs, Jr., N. van Rooijen, and I. L. Weissman. 2009. CD47 is an adverse prognostic factor and therapeutic antibody target on human acute myeloid leukemia stem cells. *Cell* 138 (2): 286–299.

Majumder, S., ed. 2009. *Stem Cells and Cancer.* New York: Springer.

Makino, S. 1956. Further evidence favoring the concept of the stem cell in ascites tumors of rats. *Annals of the New York Academy of Sciences* 63 (5): 818–830.

———. 1959. The role of tumor stem-cells in regrowth of the tumor following drastic applications. *Acta—Unio Internationalis Contra Cancrum* 15 (1): 196–198.

Makino, S., and K. Kano. 1953. Cytological studies of tumors. IX. Characteristic chromosome individuality in tumor strain-cells in ascites tumors of rats. *Journal of the National Cancer Institute* 13 (5): 1213–1235.

Malanchi, I., H. Peinado, D. Kassen, T. Hussenet, D. Metzger, P. Chambon, M. Huber, D. Hohl, A. Cano, W. Birchmeier, and J. Huelsken. 2008. Cutaneous cancer stem cell maintenance is dependent on beta-catenin signalling. *Nature* 452 (7187): 650–653.

Malanchi, I., A. Santamaria-Martinez, E. Susanto, H. Peng, H. A. Lehr, J. F. Delaloye, and J. Huelsken. 2012. Interactions between cancer stem cells and their niche govern metastatic colonization. *Nature* 481 (7379): 85–89.

Malassez, L.-C. 1885a. Sur l'existence d'amas épithéliaux autour de la racine des dents chez l'homme adulte et à l'état normal (débris épithéliaux paradentaires). *Archives de Physiologie Normale et pathologique* 3 (5): 129–148.

——. 1885b. Sur le rôle des débris épithéliaux paradentaires. *Archives de Physiologie Normale et pathologique* 3 (5): 309–340.

Mani, S. A., W. Guo, M. J. Liao, E. N. Eaton, A. Ayyanan, A. Y. Zhou, M. Brooks, F. Reinhard, C. C. Zhang, M. Shipitsin, L. L. Campbell, K. Polyak, C. Brisken, J. Yang, and R. A. Weinberg. 2008. The epithelial-mesenchymal transition generates cells with properties of stem cells. *Cell* 133 (4): 704–715.

Marjanovic, N. D., R. A. Weinberg, and C. L. Chaffer. 2013. Cell plasticity and heterogeneity in cancer. *Clinical Chemistry* 59 (1): 168–179.

Martin, B. L., and D. Kimelman. 2009. Wnt signaling and the evolution of embryonic posterior development. *Current Biology* 19 (5): R215–R219.

Martin, G. R. 1975. Teratocarcinomas as a model system for the study of embryogenesis and neoplasia. *Cell* 5 (3): 229–243.

——. 1981. Isolation of a pluripotent cell line from early mouse embryos cultured in medium conditioned by teratocarcinoma stem cells. *Proceedings of the National Academy of Sciences USA* 78 (12): 7634–7638.

Matos Rojas, I. A., D. Bertholdo, and M. Castillo. 2012. Stem cells: Implications in the development of brain tumors. *Radiología* 54 (3): 221–230.

Mauer, A. M., and V. Fisher. 1966. Characteristics of cell proliferation in four patients with untreated acute leukemia. *Blood* 28 (3): 428–445.

McAuliffe, S. M., S. L. Morgan, G. A. Wyant, L. T. Tran, K. W. Muto, Y. S. Chen, K. T. Chin, J. C. Partridge, B. B. Poole, K.-H. Cheng, J. Daggett, K. Cullen, E. Kantoff, K. Hasselbatt, J. Berkowitz, M. G. Muto, R. S. Berkowitz, J. C. Aster, U. A. Matulonis, and D. M. Dinulescu. 2012. Targeting Notch, a key pathway for ovarian cancer stem cells, sensitizes tumors to platinum therapy. *Proceedings of the National Academy of Sciences USA* 109 (43): E2939–E2948.

McCulloch, E. A., and J. E. Till. 1981. Blast cells in acute myeloblastic leukemia: A model. *Blood Cells* 7 (1): 63–77.

McCulloch, E. A., J. E. Till, and L. Siminovitch. 1965. The role of independent and dependent stem cells in the control of hemopoietic and immunologic responses. *Wistar Institute Symposium Monograph* 4: 61–68.

McCune, J. M., R. Namikawa, H. Kaneshima, L. D. Shultz, M. Lieberman, and I. L. Weissman. 1988. The SCID-hu mouse: Murine model for the analysis of human hematolymphoid differentiation and function. *Science* 241 (4873): 1632-1639.

McKenzie, J. L., O. I. Gan, M. Doedens, and J. E. Dick. 2005. Human short-term repopulating stem cells are efficiently detected following intrafemoral transplantation into NOD/SCID recipients depleted of CD122+ cells. *Blood* 106 (4): 1259-1261.

Medema, J. P., and L. Vermeulen. 2011. Microenvironmental regulation of stem cells in intestinal homeostasis and cancer. *Nature* 474 (7351): 318-326.

Mellor, D. H. 1977. Natural kinds. *British Journal for the Philosophy of Science* 28: 299-331.

Merlevede, J., N. Droin, T. Qin, K. Meldi, K. Yoshida, M. Morabito, E. Chautard, D. Auboeuf, P. Fenaux, T. Braun, R. Itzykson, S. de Botton, B. Quesnel, T. Commes, E. Jourdan, W. Vainchenker, O. Bernard, N. Pata-Merci, S. Solier, D. Selimoglu-Buet, V. Meyer, F. Artiguenave, J.-F. Deleuze, C. Preudhomme, M. R. Stratton, L. B. Alexandrov, S. Ogawa, S. Koscielny, M. Figueroa, and E. Solary. Mutation allele burden remains unchanged in chronic myelomonocytic leukemia responding to hypomethylating agents. *Nature Communications* 7: 10767.

Merlo, L. M., J. W. Pepper, B. J. Reid, and C. C. Maley. 2006. Cancer as an evolutionary and ecological process. *Nature Reviews Cancer* 6 (12): 924-935.

Metcalf, D., M. A. Moore, and N. L. Warner. 1969. Colony formation in vitro by myelomonocytic leukemic cells. *Journal of the National Cancer Institute* 43 (4): 983-1001.

Mezey, E., K. J. Chandross, G. Harta, R. A. Maki, and S. R. McKercher. 2000. Turning blood into brain: Cells bearing neuronal antigens generated in vivo from bone marrow. *Science* 290 (5497): 1779-1782.

Mikkers, H., and J. Frisen. 2005. Deconstructing stemness. *Journal of the European Molecular Biology Organization* 24 (15): 2715-2719.

Miles, J. J., S. L. Silins, A. G. Brooks, J. E. Davis, I. Misko, and S. R. Burrows. 2005. T-cell grit: Large clonal expansions of virus-specific CD8+ T cells can dominate in the peripheral circulation for at least 18 years. *Blood* 106 (13): 4412-4413.

Millikan, R. 1999. Historical kinds and the "special sciences." *Philosophical Studies* 95: 45-65.

Mintz, B. 1961. Formation and early development of germ cells. In *Symposium on the Germ Cells and Earliest Stages of Development,* edited by S. Ranzi. Milan: Instituto Lombardo.

Mintz, B., C. Cronmiller, and R. P. Custer. 1978. Somatic cell origin of teratocarcinomas. *Proceedings of the National Academy of Sciences USA* 75 (6): 2834-2838.

Mintz, B., and K. Illmensee. 1975. Normal genetically mosaic mice produced from malignant teratocarcinoma cells. *Proceedings of the National Academy of Sciences USA* 72 (9): 3585–3589.

Moncharmont, C., A. Levy, M. Gilormini, G. Bertrand, C. Chargari, G. Alphonse, D. Ardail, C. Rodriguez-Lafrasse, and N. Magne. 2012. Targeting a cornerstone of radiation resistance: Cancer stem cell. *Cancer Letters* 322 (2): 139–147.

Moore, N., and S. Lyle. 2011. Quiescent, slow-cycling stem cell populations in cancer: A review of the evidence and discussion of significance. *Journal of Oncology* 2011: 10.1155/2011/396076.

Morange, M. 2006. What history tells us VII. Twenty-five years ago: The production of mouse embryonic stem cells. *Journal of Biosciences* 31 (5): 537–541.

———. 2014. The history of stem cells. In *Stem Cells: From Basic Research to Therapy*, edited by F. Calegari and C. Vaskow. Boca Raton, FL: CRC Press.

Morel, A. P., M. Lievre, C. Thomas, G. Hinkal, S. Ansieau, and A. Puisieux. 2008. Generation of breast cancer stem cells through epithelial-mesenchymal transition. *Public Library of Science One* 3 (8): e2888.

Morgan, T. H. 1901. *Regeneration*. New York: Macmillan.

Morrison, S. J., H. D. Hemmati, A. M. Wandycz, and I. L. Weissman. 1995. The purification and characterization of fetal liver hematopoietic stem cells. *Proceedings of the National Academy of Sciences USA* 92 (22): 10302–10306.

Morrison, S. J., and I. L. Weissman. 1994. The long-term repopulating subset of hematopoietic stem cells is deterministic and isolatable by phenotype. *Immunity* 1 (8): 661–673.

Mosier, D. E., R. J. Gulizia, S. M. Baird, and D. B. Wilson. 1988. Transfer of a functional human immune system to mice with severe combined immunodeficiency. *Nature* 335 (6187): 256–259.

Mount Sinai Medical Center. 2012. Chemotherapy-resistant cancer stem cell could be "Achilles' heel" of cancer. *ScienceDaily*, September 10: www.science daily.com/releases/2012/09/120910122114.htm.

Müller, J. P. 1838. *Ueber den Feinern Bau der krankhaften Geschwuelste*. Berlin: G. Reimer.

Muller-Sieburg, C. E., R. H. Cho, L. Karlsson, J. F. Huang, and H. B. Sieburg. 2004. Myeloid-biased hematopoietic stem cells have extensive self-renewal capacity but generate diminished lymphoid progeny with impaired IL-7 responsiveness. *Blood* 103 (11): 4111–4118.

Murray, R. F., J. Hobbs, and B. Payne. 1971. Possible clonal origin of common warts (Verruca vulgaris). *Nature* 232 (5305): 51–52.

Naik, S. H., L. Perie, E. Swart, C. Gerlach, N. van Rooij, R. J. de Boer, and T. N. Schumacher. 2013. Diverse and heritable lineage imprinting of early haematopoietic progenitors. *Nature* 496 (7444): 229–232.

Naik, S. H., T. N. Schumacher, and L. Perié. 2014. Cellular barcoding: A technical appraisal. *Experimental Hematology* 42 (8): 598–608.

Najfeld, V. 1998. Clonal origin of leukemia—revisited. A tribute to Philip J Fialkow, MD. *Leukemia* 12 (2): 106–107.

Nasr, R., M. C. Guillemin, O. Ferhi, H. Soilihi, L. Peres, C. Berthier, P. Rousselot, M. Robledo-Sarmiento, V. Lallemand-Breitenbach, B. Gourmel, D. Vitoux, P. P. Pandolfi, C. Rochette-Egly, J. Zhu, and H. de Thé. 2008. Eradication of acute promyelocytic leukemia-initiating cells through PML-RARA degradation. *Nature Medicine* 14 (12): 1333–1342.

Neander, K. 1991. The teleological notion of "function." *Australasian Journal of Philosophy* 69 (4): 454–468.

Nishikawa, S., and M. Osawa. 2007. Generating quiescent stem cells. *Pigment Cell and Melanoma Research* 20 (4): 263–270.

Nobel Prize. 2012. The 2012 Nobel Prize in Physiology or Medicine—Press Release. Nobelprize.org. Nobel Media AB 2014. Web. May 20, 2015: http://www.nobelprize.org/nobel_prizes/medicine/laureates/2012/press.html.

Notta, F., S. Doulatov, and J. E. Dick. 2010. Engraftment of human hematopoietic stem cells is more efficient in female NOD/SCID/IL-2Rgc-null recipients. *Blood* 115 (18): 3704–3707.

Notta, F., C. G. Mullighan, J. C. Wang, A. Poeppl, S. Doulatov, L. A. Phillips, J. Ma, M. D. Minden, J. R. Downing, and J. E. Dick. 2011. Evolution of human BCR-ABL1 lymphoblastic leukaemia-initiating cells. *Nature* 469 (7330): 362–367.

Nowell, P. C. 1976. The clonal evolution of tumor cell populations. *Science* 194 (4260): 23–28.

Nye, H. L., J. A. Cameron, E. A. Chernoff, and D. L. Stocum. 2003. Regeneration of the urodele limb: A review. *Developmental Dynamics* 226 (2): 280–294.

Obokata, H., Y. Sasai, H. Niwa, M. Kadota, M. Andrabi, N. Takata, M. Tokoro, Y. Terashita, S. Yonemura, C. A. Vacanti, and T. Wakayama. 2014a. Bidirectional developmental potential in reprogrammed cells with acquired pluripotency. *Nature* 505 (7485): 676–680.

Obokata, H., T. Wakayama, Y. Sasai, K. Kojima, M. P. Vacanti, H. Niwa, M. Yamato, and C. A. Vacanti. 2014b. Stimulus-triggered fate conversion of somatic cells into pluripotency. *Nature* 505 (7485): 641–647.

O'Brien, C. A., A. Pollett, S. Gallinger and J. E. Dick. 2007. A human colon cancer cell capable of initiating tumour growth in immunodeficient mice. *Nature* 445 (7123): 106–110.

Odoux, C., H. Fohrer, T. Hoppo, L. Guzik, D. B. Stolz, D. W. Lewis, S. M. Gollin, T. C. Gamblin, D. A. Geller, and E. Lagasse. 2008. A Stochastic Model for Cancer Stem Cell Origin in Metastatic Colon Cancer. *Cancer Research* 68 (17): 6932–6941.

Okita, K., T. Ichisaka, and S. Yamanaka. 2007. Generation of germline-competent induced pluripotent stem cells. *Nature* 448 (7151): 313–317.

Oliver, M., and E. Scott. 1934. Adamantinoma or ameloblastoma of the hypophyseal-duct region. *American Journal of Cancer* 21 (3): 501–516.

O'Malley, M. A., and J. Dupré. 2005. Fundamental issues in systems biology. *Bioessays* 27 (12): 1270–1276.

Orkin, S. H., and L. I. Zon. 2002. Hematopoiesis and stem cells: Plasticity versus developmental heterogeneity. *Nature Immunology* 3 (4): 323–328.

Orlic, D., J. Kajstura, S. Chimenti, I. Jakoniuk, S. M. Anderson, B. Li, J. Pickel, R. McKay, B. Nadal-Ginard, D. M. Bodine, A. Leri, and P. Anversa. 2001. Bone marrow cells regenerate infarcted myocardium. *Nature* 410 (6829): 701–705.

Osafune, K., L. Caron, M. Borowiak, R. J. Martinez, C. S. Fitz-Gerald, Y. Sato, C. A. Cowan, K. R. Chien, and D. A. Melton. 2008. Marked differences in differentiation propensity among human embryonic stem cell lines. *Nature Biotechnology* 26 (3): 313–315.

Papaioannou, V. E., M. W. McBurney, R. L. Gardner, and M. J. Evans. 1975. Fate of teratocarcinoma cells injected into early mouse embryos. *Nature* 258 (5530): 70–73.

Pardal, R., M. F. Clarke, and S. J. Morrison. 2003. Applying the principles of stem-cell biology to cancer. *Nature Reviews Cancer* 3 (12): 895–902.

Park, C. H., D. E. Bergsagel, and E. A. McCulloch. 1971. Mouse myeloma tumor stem cells: A primary cell culture assay. *Journal of the National Cancer Institute* 46 (2): 411–422.

Passegue, E., C. H. Jamieson, L. E. Ailles, and I. L. Weissman. 2003. Normal and leukemic hematopoiesis: Are leukemias a stem cell disorder or a reacquisition of stem cell characteristics? *Proceedings of the National Academy of Sciences USA* 100 (Suppl 1): 11842–11849.

Pattabiraman, D. R., and R. A. Weinberg. 2014. Tackling the cancer stem cells—what challenges do they pose? *Nature Reviews Drug Discovery* 13 (7): 497–512.

Perié, L., Philip D. Hodgkin, Shalin H. Naik, Ton N. Schumacher, Rob J. de Boer, and Ken R. Duffy. 2014. Determining lineage pathways from cellular barcoding experiments. *Cell Reports* 6 (4): 617–624.

Petersen, B. E., W. C. Bowen, K. D. Patrene, W. M. Mars, A. K. Sullivan, N. Murase, S. S. Boggs, J. S. Greenberger, and J. P. Goff. 1999. Bone marrow as a potential source of hepatic oval cells. *Science* 284 (5417): 1168–1170.

Petit, I., M. Szyper-Kravitz, A. Nagler, M. Lahav, A. Peled, L. Habler, T. Ponomaryov, R. S. Taichman, F. Arenzana-Seisdedos, N. Fujii, J. Sandbank, D. Zipori, and T. Lapidot. 2002. G-CSF induces stem cell mobilization by decreasing bone marrow SDF-1 and up-regulating CXCR4. *Nature Immunology* 3 (7): 687–694.

Peyron, A. 1939. Faits nouveaux relatifs à l'origine et à l'histogenèse des embryomes. *Bulletin du Cancer* 28 (44): 658–681.

Piccirillo, S. G., S. Colman, N. E. Potter, F. W. van Delft, S. Lillis, M. J. Carnicer, L. Kearney, C. Watts, and M. Greaves. 2015. Genetic and functional diversity of propagating cells in glioblastoma. *Stem Cell Reports* 4 (1): 7–15.

Piccirillo, S. G., R. Combi, L. Cajola, A. Patrizi, S. Redaelli, A. Bentivegna, S. Baronchelli, G. Maira, B. Pollo, A. Mangiola, F. DiMeco, L. Dalpra, and A. L. Vescovi. 2009. Distinct pools of cancer stem-like cells coexist within human glioblastomas and display different tumorigenicity and independent genomic evolution. *Oncogene* 28 (15): 1807–1811.

Pierce, G. B. 1967. Teratocarcinoma: Model for a developmental concept of cancer. *Current Topics in Developmental Biology* 2: 223–246.

——. 1970. Differentiation of normal and malignant cells. *Federation Proceedings* 29 (3): 1248–1254.

——. 1974. Neoplasms, differentiations and mutations. *American Journal of Pathology* 77 (1): 103–118.

——. 1975. Teratocarcinoma: Introduction and perspectives. In *Teratomas and Differentiation,* edited by M. Sherman and D. Solter. London: Academic Press.

——. 1977a. Neoplastic stem cells. *Advances in Pathobiology* (6): 141–152.

——. 1977b. Relationship between differentiation and carcinogenesis. *Journal of Toxicology and Environmental Health* 2 (6): 1335–1342.

Pierce, G. B., and F. J. Dixon Jr. 1959a. Testicular teratomas. I. Demonstration of teratogenesis by metamorphosis of multipotential cells. *Cancer* 12 (3): 573–583.

——. 1959b. Testicular teratomas. II. Teratocarcinoma as an ascitic tumor. *Cancer* 12 (3): 584–589.

Pierce, G. B., F. J. Dixon Jr., and E. L. Verney. 1960. Teratocarcinogenic and tissue-forming potentials of the cell types comprising neoplastic embryoid bodies. *Laboratory Investigation* 9: 583–602.

Pierce, G. B., and L. D. Johnson. 1971. Differentiation and cancer. *In Vitro* 7 (3): 140–145.

Pierce, G. B., P. K. Nakane, A. Martinez-Hernandez, and J. M. Ward. 1977. Ultrastructural comparison of differentiation of stem cells of murine adenocarcinomas of colon and breast with their normal counterparts. *Journal of the National Cancer Institute* 58 (5): 1329–1345.

Pierce, G. B., R. Shikes, and L. M. Fink. 1978. *Cancer: A Problem of Developmental Biology.* Englewood Cliffs, NJ: Prentice Hall.

Pierce, G. B., and E. L. Verney. 1961. An in vitro and in vivo study of differentiation in teratocarcinomas. *Cancer* 14: 1017–1029.

Pisco, A. O., and S. Huang. 2015. Non-genetic cancer cell plasticity and therapy-induced stemness in tumour relapse: "What does not kill me strengthens me." *British Journal of Cancer* 112 (11): 1725–1732.

Pittenger, M. F., A. M. Mackay, S. C. Beck, R. K. Jaiswal, R. Douglas, J. D. Mosca, M. A. Moorman, D. W. Simonetti, S. Craig, and D. R. Marshak. 1999. Multilineage potential of adult human mesenchymal stem cells. *Science* 284 (5411): 143–147.

Plaks, V., N. Kong, and Z. Werb. 2015. The cancer stem cell niche: How essential is the niche in regulating stemness of tumor cells? *Cell Stem Cell* 16 (3): 225–238.

Polyak, K. 2007. Breast cancer: Origins and evolution. *Journal of Clinical Investigation* 117 (11): 3155–3163.

Polyak, K., and W. C. Hahn. 2006. Roots and stems: Stem cells in cancer. *Nature Medicine* 12 (3): 296–300.

Poss, K. D. 2010. Advances in understanding tissue regenerative capacity and mechanisms in animals. *Nature Reviews Genetics* 11 (10): 710–722.

Potten, C. S., and M. Loeffler. 1990. Stem cells: Attributes, cycles, spirals, pitfalls and uncertainties. Lessons for and from the crypt. *Development* 110 (4): 1001–1020.

de Poulton Nicholson, G. W. 1929. The histogeny of teratomata. *Journal of Pathology and Bacteriology* 32 (3): 365–386.

Pradeu, T. 2012. *The Limits of the Self: Immunology and Biological Identity.* New York: Oxford University Press.

Priller, J., D. A. Persons, F. F. Klett, G. Kempermann, G. W. Kreutzberg, and U. Dirnagl. 2001. Neogenesis of cerebellar Purkinje neurons from gene-marked bone marrow cells in vivo. *Journal of Cell Biology* 155 (5): 733–738.

Prince, M. E., R. Sivanandan, A. Kaczorowski, G. T. Wolf, M. J. Kaplan, P. Dalerba, I. L. Weissman, M. F. Clarke, and L. E. Ailles. 2007. Identification of a subpopulation of cells with cancer stem cell properties in head and neck squamous cell carcinoma. *Proceedings of the National Academy of Sciences USA* 104 (3): 973–978.

Qian, H., N. Buza-Vidas, C. D. Hyland, C. T. Jensen, J. Antonchuk, R. Mansson, L. A. Thoren, M. Ekblom, W. S. Alexander, and S. E. W. Jacobsen. 2007. Critical role of thrombopoietin in maintaining adult quiescent hematopoietic stem cells. *Cell Stem Cell* 1 (6): 671–684.

Quesenberry, P. J., G. Colvin, G. Dooner, M. Dooner, J. M. Aliotta, and K. Johnson. 2007. The stem cell continuum: Cell cycle, injury, and phenotype lability. *Annals of the New York Academy of Sciences* 1106: 20–29.

Quesenberry, P. J., G. A. Colvin, and J. F. Lambert. 2002. The chiaroscuro stem cell: A unified stem cell theory. *Blood* 100 (13): 4266–4271.

Quintana, E., M. Shackleton, M. S. Sabel, D. R. Fullen, T. M. Johnson, and S. J. Morrison. 2008. Efficient tumour formation by single human melanoma cells. *Nature* 456 (7222): 593–598.

Rader, K. 2004. *Making Mice: Standardizing Animals for American Biomedical Research, 1900–1955.* Princeton, NJ: Princeton University Press.

Rajasekhar, V. K., ed. 2014. *Cancer Stem Cells.* Hoboken, NJ: Wiley-Blackwell.

Rak, J. 2006. Is cancer stem cell a cell, or a multicellular unit capable of inducing angiogenesis? *Medical Hypotheses* 66 (3): 601–604.

Ramachandran, R., B. V. Fausett, and D. Goldman. 2010. Ascl1a regulates Muller glia dedifferentiation and retinal regeneration through a Lin-28-dependent, let-7 microRNA signalling pathway. *Nature Cell Biology* 12 (11): 1101–1107.

Ramalho-Santos, M., and H. Willenbring. 2007. On the origin of the term "stem cell." *Cell Stem Cell* 1 (1): 35–38.

Ramalho-Santos, M., S. Yoon, Y. Matsuzaki, R. C. Mulligan, and D. A. Melton. 2002. "Stemness": Transcriptional profiling of embryonic and adult stem cells. *Science* 298 (5593): 597–600.

Rankin, J. O. 1939. Adamantinoma of the tibia. *Journal of Bone and Joint Surgery* 21 (2): 425–432.

Rapp, U. R., F. Ceteci, and R. Schreck. 2008. Oncogene-induced plasticity and cancer stem cells. *Cell Cycle* 7 (1): 45–51.

Raskind, W. H., L. Steinmann, and V. Najfeld. 1998. Clonal development of myeloproliferative disorders: Clues to hematopoietic differentiation and multistep pathogenesis of cancer. *Leukemia* 12 (2): 108–116.

Rastrick, J. M. 1969. A method for the positive identification of erythropoietic cells in chromosome preparations of bone marrow. *British Journal of Haematology* 16 (1): 185–191.

Rather, L. J. 1978. *The Genesis of Cancer: A Study in the History of Ideas.* Baltimore: Johns Hopkins University Press.

Remak, R. 1854. Ein Beitrag zur Entwickelungsgeschichte der krebshaften Geschwuelste. *Deutsche Klinik:* 170–175.

——. 1855. *Untersuchungen über die Entwickelung der Wilbelthiere.* Berlin: R. Reimer.

Restifo, N. P., M. E. Dudley, and S. A. Rosenberg. 2012. Adoptive immunotherapy for cancer: harnessing the T cell response. *Nature Reviews Immunology* 12 (4): 269–281.

Reya, T., A. W. Duncan, L. Ailles, J. Domen, D. C. Scherer, K. Willert, L. Hintz, R. Nusse, and I. L. Weissman. 2003. A role for Wnt signalling in self-renewal of haematopoietic stem cells. *Nature* 423 (6938): 409–414.

Reya, T., S. J. Morrison, M. F. Clarke, and I. L. Weissman. 2001. Stem cells, cancer, and cancer stem cells. *Nature* 414 (6859): 105–111.

Rheinberger, H.-J. 2000. Gene concepts: Fragments from the perspective of molecular biology. In *The Concept of the Gene in Development and Evolution: Historical and Epistemological Perspectives,* edited by P. Beurton, R. Falk, and H.-J. Rheinberger. Cambridge: Cambridge University Press.

Ricci-Vitiani, L., A. Pagliuca, E. Palio, A. Zeuner, and R. De Maria. 2008. Colon cancer stem cells. *Gut* 57 (4): 538–548.

Ricciardi, Michael. 2012. Cancer treatment breakthrough: A single antibody drug found to shrink or halt "all tumors." *Planetsave,* March 27: http://planetsave

.com/2012/03/27/cancer-treatment-breakthrough-a-single-antibody
-drug-found-to-shrink-or-halt-all-tumors/#95bT7fqUMi7avxQU.99.

Rice, K. L., and H. de Thé. 2014. The acute promyelocytic leukaemia success story: Curing leukaemia through targeted therapies. *Journal of Internal Medicine* 276 (1): 61–70.

Richter, C. S. 1930. Ein Fall von adamantinomartiger Geschwulst des Schien-beins. *Zeitschrift für Krebsforschung* 32 (1–2): 273–279.

Robert, J. S. 2004. Model systems in stem cell biology. *Bioessays* 26 (9): 1005–1012.

Roesch, A., M. Fukunaga-Kalabis, E. C. Schmidt, S. E. Zabierowski, P. A. Braf-ford, A. Vultur, D. Basu, P. Gimotty, T. Vogt, and M. Herlyn. 2010. A temporarily distinct subpopulation of slow-cycling melanoma cells is required for continuous tumor growth. *Cell* 141 (4): 583–594.

Rosai, J. 1969. Adamantinoma of the tibia: Electron microscopic evidence of its epithelial origin. *American Journal of Clinical Pathology* 51 (6): 786–792.

Rosai, J., and G. S. Pincus. 1982. Immunohistochemical demonstration of epi-thelial differentiation in adamantinoma of the tibia. *American Journal of Surgical Pathology* 6 (5): 427–434.

Rosenberg, S. A., J. C. Yang, R. M. Sherry, U. S. Kammula, M. S. Hughes, G. Q. Phan, D. E. Citrin, N. P. Restifo, P. F. Robbins, J. R. Wunderlich, K. E. Morton, C. M. Laurencot, S. M. Steinberg, D. E. White, and M. E. Dudley. 2011. Durable complete responses in heavily pretreated patients with metastatic melanoma using T-cell transfer immunotherapy. *Clinical Cancer Research* 17 (13): 4550–4557.

Rousselot, P., F. Huguet, D. Rea, L. Legros, J. M. Cayuela, O. Maarek, O. Blan-chet, G. Marit, E. Gluckman, J. Reiffers, M. Gardembas, and F. X. Mahon. 2007. Imatinib mesylate discontinuation in patients with chronic my-elogenous leukemia in complete molecular remission for more than 2 years. *Blood* 109 (1): 58–60.

Safa, A. R., M. R. Saadatzadeh, A. A. Cohen-Gadol, K. E. Pollok, and K. Bijangi-Vishehsaraei. 2015. Glioblastoma stem cells (GSCs) epigenetic plasticity and interconversion between differentiated non-GSCs and GSCs. *Genes & Diseases* 2 (2): 152–163.

Saito, Y., N. Uchida, S. Tanaka, N. Suzuki, M. Tomizawa-Murasawa, A. Sone, Y. Najima, S. Takagi, Y. Aoki, A. Wake, S. Taniguchi, L. D. Shultz, and F. Ishikawa. 2010. Induction of cell cycle entry eliminates human leu-kemia stem cells in a mouse model of AML. *Nature Biotechnology* 28 (3): 275–280.

Salsbury, A. J. 1975. The significance of the circulating cancer cell. *Cancer Treat-ment Reviews* 2 (1): 55–72.

Sato, H. 1952. On the chromosomes of Yoshida sarcoma; Studies with tumor cells proliferated in the peritoneal cavity of the rat transplanted with a single cell. *Gan* 43 (1): 1–16.

Satoh, A., S. V. Bryant, and D. M. Gardiner. 2008a. Regulation of dermal fibroblast dedifferentiation and redifferentiation during wound healing and limb regeneration in the Axolotl. *Development Growth and Differentiation* 50 (9): 743–754.

Satoh, A., G. M. Graham, S. V. Bryant, and D. M. Gardiner. 2008b. Neurotrophic regulation of epidermal dedifferentiation during wound healing and limb regeneration in the axolotl (Ambystoma mexicanum). *Developmental Biology* 319 (2): 321–335.

Scatena, R., A. Mordente, and B. Giardina, eds. 2011. *Advances in Cancer Stem Cell Biology*. New York: Springer.

Scheel, C., E. N. Eaton, S. H. Li, C. L. Chaffer, F. Reinhardt, K. J. Kah, G. Bell, W. Guo, J. Rubin, A. L. Richardson, and R. A. Weinberg. 2011. Paracrine and autocrine signals induce and maintain mesenchymal and stem cell states in the breast. *Cell* 145 (6): 926–940.

Schepers, K., T. B. Campbell, and E. Passegué. 2015. Normal and leukemic stem cell niches: Insights and therapeutic opportunities. *Cell Stem Cell* 16 (3): 254–267.

Schleiden, M. J. 1838. Beiträge zur Phytogenesis. *Archiv für Anatomie, Physiologie und wissenschaftliche Medicin:* 137–176.

Schofield, R. 1978. The relationship between the spleen colony-forming cell and the haemopoietic stem cell. *Blood Cells* 4 (1–2): 7–25.

———. 1983. The stem cell system. *Biomedicine and Pharmacotherapy* 37 (8): 375–380.

Schulenburg, C. A. 1951. Adamantinoma. *Annals of the Royal College of Surgeons of England* 8 (5): 329–353.

Schwann, T. 1839. *Mikroskopische Untersuchungen über die Übereinstimmung in der Struktur und dem Wachstum der Thiere und Pflanzen*. Berlin: Sander'schen Buchhandlung.

Schwendemann, J., C. Choi, V. Schirrmacher, and P. Beckhove. 2005. Dynamic differentiation of activated human peripheral blood CD8+ and CD4+ effector memory T cells. *Journal of Immunology* 175 (3): 1433–1439.

Seaberg, R. M., and D. van der Kooy. 2003. Stem and progenitor cells: The premature desertion of rigorous definitions. *Trends in Neurosciences* 26 (3): 125–131.

Seery, J. P., and F. M. Watt. 2000. Asymmetric stem-cell divisions define the architecture of human oesophageal epithelium. *Current Biology* 10 (22): 1447–1450.

Sell, S. 2004. Stem cell origin of cancer and differentiation therapy. *Critical Reviews in Oncology/Hematology* 51 (1): 1–28.

———. 2006. Cancer stem cells and differentiation therapy. *Tumor Biology* 27 (2): 59–70.

———. 2009. History of cancer stem cells. In *Regulatory Networks in Stem Cells*, edited by V. K. Rajasekhar and M. C. Vemuri. N.p.: Humana Press.

Sell, S., and G. B. Pierce. 1994. Maturation arrest of stem cell differentiation is a common pathway for the cellular origin of teratocarcinomas and epithelial cancers. *Laboratory Investigation* 70 (1): 6–22.

Shen, C. N., M. E. Horb, J. M. Slack, and D. Tosh. 2003. Transdifferentiation of pancreas to liver. *Mechanisms of Development* 120 (1): 107–116.

Sheng, X. R., C. M. Brawley, and E. L. Matunis. 2009. Dedifferentiating spermatogonia outcompete somatic stem cells for niche occupancy in the Drosophila testis. *Cell Stem Cell* 5 (2): 191–203.

Sherman, M. I. 1975a. The culture of cells derived from mouse blastocysts. *Cell* 5 (4): 343–349.

———. 1975b. Long term culture of cells derived from mouse blastocysts. *Differentiation* 3 (1–3): 51–67.

Shi, Y., C. Desponts, J. T. Do, H. S. Hahm, H. R. Schöler, and S. Ding. 2008. Induction of pluripotent stem cells from mouse embryonic fibroblasts by oct4 and klf4 with small-molecule compounds. *Cell Stem Cell* 3 (5): 568–574.

Shlush, L. I., S. Zandi, A. Mitchell, W. C. Chen, J. M. Brandwein, V. Gupta, J. A. Kennedy, A. D. Schimmer, A. C. Schuh, K. W. Yee, J. L. McLeod, M. Doedens, J. J. Medeiros, R. Marke, H. J. Kim, K. Lee, J. D. McPherson, T. J. Hudson, H. P.-L. G. P. Consortium, A. M. Brown, F. Yousif, Q. M. Trinh, L. D. Stein, M. D. Minden, J. C. Wang, and J. E. Dick. 2014. Identification of pre-leukaemic haematopoietic stem cells in acute leukaemia. *Nature* 506 (7488): 328–333.

Shostak, S. 2006. (Re)defining stem cells. *Bioessays* 28 (3): 301–308.

———, ed. 2011. *Cancer Stem Cells: The Cutting Edge.* Rijeka: InTech.

Sieweke, M. H., and J. E. Allen. 2013. Beyond stem cells: Self-renewal of differentiated macrophages. *Science* 342 (6161): 10.1126/science.1242974.

Silva, J., O. Barrandon, J. Nichols, J. Kawaguchi, T. W. Theunissen, and A. Smith. 2008. Promotion of reprogramming to ground state pluripotency by signal inhibition. *Public Library of Science Biology* 6 (10): e253.

Simson, L. R., I. Lampe, and M. R. Abell. 1968. Suprasellar germinomas. *Cancer* 22 (3): 533–544.

Singh, S. K., N.-M. Chen, E. Hessmann, J. Siveke, M. Lahmann, G. Singh, N. Voelker, S. Vogt, I. Esposito, A. Schmidt, C. Brendel, T. Stiewe, J. Gaedcke, M. Mernberger, H. C. Crawford, W. R. Bamlet, J.-S. Zhang, X.-K. Li, T. C. Smyrk, D. D. Billadeau, M. Hebrok, A. Neesse, A. Koenig, and V. Ellenrieder. 2015. Antithetical NFATc1-Sox2 and p53-miR200 signaling networks govern pancreatic cancer cell plasticity. *Journal of the European Molecular Biology Organization* 34 (4): 517–530.

Singh, S. K., I. D. Clarke, M. Terasaki, V. E. Bonn, C. Hawkins, J. Squire, and P. B. Dirks. 2003. Identification of a cancer stem cell in human brain tumors. *Cancer Research* 63 (18): 5821–5828.

Singh, S. K., C. Hawkins, I. D. Clarke, J. A. Squire, J. Bayani, T. Hide, R. M. Henkelman, M. D. Cusimano, and P. B. Dirks. 2004. Identification of human brain tumour initiating cells. *Nature* 432 (7015): 396–401.

Sneddon, J. B., and Z. Werb. 2007. Location, location, location: the cancer stem cell niche. *Cell Stem Cell* 1 (6): 607–611.

Solovitch, Sara. 2008. Cancer research nets $1.4B deal for OncoMed. *San Francisco Business Times,* January 13: http://www.bizjournals.com/sanfrancisco/stories/2008/01/14/story13.html?page=all.

Song, X., and T. Xie. 2002. DE-cadherin-mediated cell adhesion is essential for maintaining somatic stem cells in the Drosophila ovary. *Proceedings of the National Academy of Sciences USA* 99 (23): 14813–14818.

Song, X., C. H. Zhu, C. Doan, and T. Xie. 2002. Germline stem cells anchored by adherens junctions in the Drosophila ovary niches. *Science* 296 (5574): 1855–1857.

Sourisseau, T., K. A. Hassan, I. Wistuba, F. Penault-Llorca, J. Adam, E. Deutsch, and J. C. Soria. 2014. Lung cancer stem cell: Fancy conceptual model of tumor biology or cornerstone of a forthcoming therapeutic breakthrough? *Journal of Thoracic Oncology* 9 (1): 7–17.

Southam, C. M., and A. Brunschwig. 1961. Quantitative studies of autotransplantation of human cancer. Preliminary report. *Cancer* 14 (5): 971–978.

Stadtfeld, M., and K. Hochedlinger. 2010. Induced pluripotency: History, mechanisms, and applications. *Genes and Development* 24 (20): 2239–2263.

Stadtfeld, M., M. Nagaya, J. Utikal, G. Weir, and K. Hochedlinger. 2008. Induced pluripotent stem cells generated without viral integration. *Science* 322 (5903): 945–949.

Staerk, J., M. M. Dawlaty, Q. Gao, D. Maetzel, J. Hanna, C. A. Sommer, G. Mostoslavsky, and R. Jaenisch. 2010. Reprogramming of human peripheral blood cells to induced pluripotent stem cells. *Cell Stem Cell* 7 (1): 20–24.

Stegmeier, F., M. Warmuth, W. R. Sellers, and M. Dorsch. 2010. Targeted cancer therapies in the twenty-first century: Lessons from imatinib. *Clinical Pharmacology and Therapeutics* 87 (5): 543–552.

Stevens, L. C. 1959. Embryology of testicular teratomas in strain 129 mice. *Journal of the National Cancer Institute* 23: 1249–1295.

———. 1960. Embryonic potency of embryoid bodies derived from a transplantable testicular teratoma of the mouse. *Developmental Biology* 2: 285–297.

———. 1964. Experimental production of testicular teratomas in mice. *Proceedings of the National Academy of Sciences USA* 52: 654–661.

———. 1967. Origin of testicular teratomas from primordial germ cells in mice. *Journal of the National Cancer Institute* 38 (4): 549–552.

———. 1968. The development of teratomas from intratesticular grafts of tubal mouse eggs. *Journal of Embryology and Experimental Morphology* 20 (3): 329–341.

———. 1970. The development of transplantable teratocarcinomas from intratesticular grafts of pre- and postimplantation mouse embryos. *Developmental Biology* 21 (3): 364–382.

Stevens, L. C., and C. C. Little. 1954. Spontaneous testicular teratomas in an inbred strain of mice. *Proceedings of the National Academy of Sciences USA* 40 (11): 1080–1087.

Strathern, M. 1988. *The Gender of the Gift: Problems with Women and Problems with Society in Melanesia.* Berkeley: University of California Press.

Stratton, M. R. 2011. Exploring the genomes of cancer cells: Progress and promise. *Science* 331 (6024): 1553–1558.

Succony, L., and S. M. Janes. 2014. Airway stem cells and lung cancer. *Quarterly Journal of Medicine* 107 (8): 607–12.

Suda, T., and F. Arai. 2008. Wnt signaling in the niche. *Cell* 132 (5): 729–730.

Suetsugu, A., M. Nagaki, H. Aoki, T. Motohashi, T. Kunisada, and H. Moriwaki. 2006. Characterization of CD133+ hepatocellular carcinoma cells as cancer stem/progenitor cells. *Biochemical and Biophysical Research Communications* 351 (4): 820–824.

Susman, W. 1932. Embryonic epithelial rests in the pituitary. *British Journal of Surgery* 19 (76): 671–676.

Szotek, P. P., R. Pieretti-Vanmarcke, P. T. Masiakos, D. M. Dinulescu, D. Connolly, R. Foster, D. Dombkowski, F. Preffer, D. T. Maclaughlin, and P. K. Donahoe. 2006. Ovarian cancer side population defines cells with stem cell-like characteristics and Mullerian Inhibiting Substance responsiveness. *Proceedings of the National Academy of Sciences USA* 103 (30): 11154–11159.

Tachibana, M., P. Amato, M. Sparman, N. M. Gutierrez, R. Tippner-Hedges, H. Ma, E. Kang, A. Fulati, H.-S. Lee, H. Sritanaudomchai, K. Masterson, J. Larson, D. Eaton, K. Sadler-Fredd, D. Battaglia, D. Lee, D. Wu, J. Jensen, P. Patton, S. Gokhale, R. L. Stouffer, D. Wolf, and S. Mitalipov. 2013. Human embryonic stem cells derived by somatic cell nuclear transfer. *Cell* 153 (6): 1228–1238.

Takahashi, K., K. Tanabe, M. Ohnuki, M. Narita, T. Ichisaka, K. Tomoda, and S. Yamanaka. 2007. Induction of pluripotent stem cells from adult human fibroblasts by defined factors. *Cell* 131 (5): 861–872.

Takahashi, K., and S. Yamanaka. 2006. Induction of pluripotent stem cells from mouse embryonic and adult fibroblast cultures by defined factors. *Cell* 126 (4): 663–676.

Takebe, N., and S. P. Ivy. 2010. Controversies in cancer stem cells: Targeting embryonic signaling pathways. *Clinical Cancer Research* 16 (12): 3106–3112.

Tallman, M. S., J. W. Andersen, C. A. Schiffer, F. R. Appelbaum, J. H. Feusner, W. G. Woods, A. Ogden, H. Weinstein, L. Shepherd, C. Willman, C. D. Bloomfield, J. M. Rowe, and P. H. Wiernik. 2002. All-trans retinoic acid in acute promyelocytic leukemia: Long-term outcome and prognostic factor analysis from the North American Intergroup protocol. *Blood* 100 (13): 4298–4302.

Talpaz, M., R. Hehlmann, A. Quintas-Cardama, J. Mercer, and J. Cortes. 2013. Re-emergence of interferon-alpha in the treatment of chronic myeloid leukemia. *Leukemia* 27 (4): 803–812.

Tan, B. T., C. Y. Park, L. E. Ailles, and I. L. Weissman. 2006. The cancer stem cell hypothesis: a work in progress. *Laboratory Investigation* 86 (12): 1203–1207.

Taussig, D. C., F. Miraki-Moud, F. Anjos-Afonso, D. J. Pearce, K. Allen, C. Ridler, D. Lillington, H. Oakervee, J. Cavenagh, S. G. Agrawal, T. A. Lister, J. G. Gribben, and D. Bonnet. 2008. Anti-CD38 antibody-mediated clearance of human repopulating cells masks the heterogeneity of leukemia-initiating cells. *Blood* 112 (3): 568–575.

Taussig, D. C., J. Vargaftig, F. Miraki-Moud, E. Griessinger, K. Sharrock, T. Luke, D. Lillington, H. Oakervee, J. Cavenagh, S. G. Agrawal, T. A. Lister, J. G. Gribben, and D. Bonnet. 2010. Leukemia-initiating cells from some acute myeloid leukemia patients with mutated nucleophosmin reside in the CD34(-) fraction. *Blood* 115 (10): 1976–1984.

Tetteh, P. W., H. F. Farin, and H. Clevers. 2015. Plasticity within stem cell hierarchies in mammalian epithelia. *Trends in Cell Biology* 25 (2): 100–108.

Theise, N. D., S. Badve, R. Saxena, O. Henegariu, S. Sell, J. M. Crawford, and D. S. Krause. 2000a. Derivation of hepatocytes from bone marrow cells in mice after radiation-induced myeloablation. *Hepatology* 31 (1): 235–240.

Theise, N. D., M. Nimmakayalu, R. Gardner, P. B. Illei, G. Morgan, L. Teperman, O. Henegariu, and D. S. Krause. 2000b. Liver from bone marrow in humans. *Hepatology* 32 (1): 11–16.

Theise, N. D., and I. Wilmut. 2003. Cell plasticity: Flexible arrangement. *Nature* 425 (6953): 21.

Therasse, P., S. G. Arbuck, E. A. Eisenhauer, J. Wanders, R. S. Kaplan, L. Rubinstein, J. Verweij, M. Van Glabbeke, A. T. van Oosterom, M. C. Christian, and S. G. Gwyther. 2000. New guidelines to evaluate the response to treatment in solid tumors. European Organization for Research and Treatment of Cancer, National Cancer Institute of the United States, National Cancer Institute of Canada. *Journal of the National Cancer Institute* 92 (3): 205–216.

Thiel, E., P. Dormer, H. Rodt, D. Huhn, M. Bauchinger, H. P. Kley, and S. Thierfelder. 1977. Quantitation of T-antigenic sites and Ig-determinants on leukemic cells by microphotometric immunoautoradiography. Proof of the clonal origin of thymus-derived lymphocytic leukemias. *Haematology and Blood Transfusion* 20: 131–145.

Thiel, E., H. Rodt, D. Huhn, and S. Thierfelder. 1976. Decrease and altered distribution of human T antigen on chronic lymphatic leukemia cells of T type, suggesting a clonal origin. *Blood* 47 (5): 723–736.

Thirant, C., B. Bessette, P. Varlet, S. Puget, J. Cadusseau, R. Tavares Sdos, J. M. Studler, D. C. Silvestre, A. Susini, C. Villa, C. Miquel, A. Bogeas, A. L. Surena, A. Dias-Morais, N. Leonard, F. Pflumio, I. Bieche, F. D. Boussin, C. Sainte-Rose, J. Grill, C. Daumas-Duport, H. Chneiweiss, and M. P. Junier. 2011. Clinical relevance of tumor cells with stem-like properties in pediatric brain tumors. *Public Library of Science One* 6 (1): e16375.

Thomson, J. A., J. Itskovitz-Eldor, S. S. Shapiro, M. A. Waknitz, J. J. Swiergiel, V. S. Marshall, and J. M. Jones. 1998. Embryonic stem cell lines derived from human blastocysts. *Science* 282 (5391): 1145–1147.

Till, J. E., and E. A. McCulloch. 1961. A direct measurement of the radiation sensitivity of normal mouse bone marrow cells. *Radiation Research* 14: 213–222.

———. 1980. Hemopoietic stem cell differentiation. *Biochimica et Biophysica Acta* 605 (4): 431–459.

Till, J. E., E. A. McCulloch, and L. Siminovitch. 1964. A stochastic model of stem cell proliferation, based on the growth of spleen colony-forming cells. *Proceedings of the National Academy of Sciences USA* 51: 29–36.

Tough, I. M., P. A. Jacobs, W. M. Court Brown, A. G. Baikie, and E. R. Williamson. 1963. Cytogenetic studies on bone-marrow in chronic myeloid leukaemia. *Lancet* 1 (7286): 844–846.

Trentin, A., C. Glavieux-Pardanaud, N. M. Le Douarin, and E. Dupin. 2004. Self-renewal capacity is a widespread property of various types of neural crest precursor cells. *Proceedings of the National Academy of Sciences USA* 101 (13): 4495–4500.

Trujillo, J. M., and S. Ohno. 1963. Chromosomal alteration of erythropoietic cells in chronic myeloid leukemia. *Acta Haematologica* 29: 311–316.

Tsukamoto, H., H. She, S. Hazra, J. Cheng, and T. Miyahara. 2006. Anti-adipogenic regulation underlies hepatic stellate cell transdifferentiation. *Journal of Gastroenterology and Hepatology* 21 (Suppl 3): S102–S105.

Tu, S.-M. 2010. *Origin of Cancers: Clinical Perspectives and Implications of a Stem-Cell Theory of Cancers.* New York: Springer.

Tulina, N., and E. Matunis. 2001. Control of stem cell self-renewal in Drosophila spermatogenesis by JAK-STAT signaling. *Science* 294 (5551): 2546–2549.

Turhan, A. G., F. M. Lemoine, C. Debert, M. L. Bonnet, C. Baillou, F. Picard, E. A. Macintyre, and B. Varet. 1995. Highly purified primitive hematopoietic stem cells are PML-RARA negative and generate nonclonal progenitors in acute promyelocytic leukemia. *Blood* 85 (8): 2154–2161.

Valent, P., D. Bonnet, R. De Maria, T. Lapidot, M. Copland, J. V. Melo, C. Chomienne, F. Ishikawa, J. J. Schuringa, G. Stassi, B. Huntly, H. Herrmann, J. Soulier, A. Roesch, G. J. Schuurhuis, S. Wohrer, M. Arock, J. Zuber, S. Cerny-Reiterer, H. E. Johnsen, M. Andreeff and C. Eaves. 2012. Cancer stem cell definitions and terminology: The devil is in the details. *Nature Reviews Cancer* 12 (11): 767–775.

van Bekkum, D. W., and S. Knaan. 1978. Similarity in morphological appearance of cycling and resting hemopoietic stem cells. *Bulletin du Cancer* 65 (4): 437–441.

Vermeulen, L., F. de Sousa e Melo, M. van der Heijden, K. Cameron, J. H. de Jong, T. Borovski, J. B. Tuynman, M. Todaro, C. Merz, H. Rodermond, M. R. Sprick, K. Kemper, D. J. Richel, G. Stassi, and J. P. Medema. 2010. Wnt activity defines colon cancer stem cells and is regulated by the microenvironment. *Nature Cell Biology* 12 (5): 468–476.

Vermeulen, L., F. de Sousa e Melo, D. J. Richel, and J. P. Medema. 2012. The developing cancer stem-cell model: Clinical challenges and opportunities. *Lancet Oncology* 13 (2): e83–e89.

Vermeulen, L., M. R. Sprick, K. Kemper, G. Stassi, and J. P. Medema. 2008. Cancer stem cells—Old concepts, new insights. *Cell Death and Differentiation* 15 (6): 947–958.

Vidal, S. J., V. Rodriguez-Bravo, M. Galsky, C. Cordon-Cardo, and J. Domingo-Domenech. 2014. Targeting cancer stem cells to suppress acquired chemotherapy resistance. *Oncogene* 33 (36): 4451–4463.

Vineis, P., and M. Berwick. 2006. The population dynamics of cancer: A Darwinian perspective. *International Journal of Epidemiology* 35 (5): 1151–1159.

Vinogradov, S., and X. Wei. 2012. Cancer stem cells and drug resistance: The potential of nanomedicine. *Nanomedicine (London)* 7 (4): 597–615.

Virchow, R. 1850. Ueber Kankroide und Papillargeschwuelste. *Verhandlungen der Physikalisch-Medizinischen Gesellschaft zu Würzburg* 1: 106–111.

——. 1855a. Cellular-Pathologie. *Archiv für pathologische Anatomie und Physiologie und für klinische Medizin* 8: 1–33.

——. 1855b. Ueber Perlgeschwuelste (Cholesteatoma Joh. Mueller's). *Archiv für pathologische Anatomie und Physiologie und für klinische Medizin* 8: 371–415.

——. 1860. *Cellular Pathology, as based upon Physiological and Pathological Histology.* 2nd ed. Translated by Frank Chance. London: John Churchill.

Visvader, J. E. 2011. Cells of origin in cancer. *Nature* 469 (7330): 314–322.

Visvader, J. E., and G. J. Lindeman. 2012. Cancer stem cells: Current status and evolving complexities. *Cell Stem Cell* 10 (6): 717–728.

Vogel, G. 2003. Stem cells. "Stemness" genes still elusive. *Science* 302 (5644): 371.

——. 2013. Human stem cells from cloning, finally. *Science* 340 (6134): 795.

Wagers, A. J., and I. L. Weissman. 2004. Plasticity of adult stem cells. *Cell* 116 (5): 639–648.

Wakao, S., M. Kitada, and M. Dezawa. 2013. The elite and stochastic model for iPS cell generation: Multilineage-differentiating stress enduring (Muse) cells are readily reprogrammable into iPS cells. *Cytometry Part A* 83A (1): 18–26.

Wakao, S., M. Kitada, Y. Kuroda, T. Shigemoto, D. Matsuse, H. Akashi, Y. Tanimura, K. Tsuchiyama, T. Kikuchi, M. Goda, T. Nakahata, Y. Fujiyoshi, and M. Dezawa. 2011. Multilineage-differentiating stress-enduring

(Muse) cells are a primary source of induced pluripotent stem cells in human fibroblasts. *Proceedings of the National Academy of Sciences USA* 108 (24): 9875–9880.

Wakayama, S., T. Kohda, H. Obokata, M. Tokoro, C. Li, Y. Terashita, E. Mizu-tani, V. T. Nguyen, S. Kishigami, F. Ishino, and T. Wakayama. 2013. Successful serial recloning in the mouse over multiple generations. *Cell Stem Cell* 12 (3): 293–297.

Wan, J., R. Ramachandran, and D. Goldman. 2012. HB-EGF is necessary and sufficient for Muller glia dedifferentiation and retina regeneration. *Developmental Cell* 22 (2): 334–347.

Wang, J. C., and J. E. Dick. 2005. Cancer stem cells: Lessons from leukemia. *Trends in Cell Biology* 15 (9): 494–501.

Wang, K., X. Wu, J. Wang, and J. Huang. 2013. Cancer stem cell theory: Therapeutic implications for nanomedicine. *International Journal of Nanomedicine* 8: 899–908.

Wang, L., P. Park, H. Zhang, F. La Marca, A. Claeson, J. Valdivia, and C.-Y. Lin. 2011. BMP-2 inhibits the tumorigenicity of cancer stem cells in human osteosarcoma OS99-1 cell line. *Cancer Biology and Therapy* 11 (5): 457–463.

Wang, Z., and G. Ouyang. 2012. Periostin: A bridge between cancer stem cells and their metastatic niche. *Cell Stem Cell* 10 (2): 111–112.

Wang, Z. Y., and Z. Chen. 2008. Acute Promyelocytic Leukemia: From Highly Fatal to Highly Curable. *Blood* 111 (5): 2505–2515.

Watt, F. M., and R. R. Driskell. 2010. The therapeutic potential of stem cells. *Philosophical transactions of the Royal Society of London. Series B, Biological Sciences* 365 (1537): 155–63.

Wei, J., M. Wunderlich, C. Fox, S. Alvarez, J. C. Cigudosa, J. S. Wilhelm, Y. Zheng, J. A. Cancelas, Y. Gu, M. Jansen, J. F. Dimartino, and J. C. Mulloy. 2008. Microenvironment determines lineage fate in a human model of MLL-AF9 leukemia. *Cancer Cell* 13 (6): 483–495.

Weissman, I. L. 2000. Stem cells: Units of development, units of regeneration, and units in evolution. *Cell* 100 (1): 157–168.

Weissman, I. L., R. Majeti, A. A. Alizadeh, and M. P. Chao. 2012. Synergistic anti-CD47 therapy for hematologic cancers. US Patent US20120282174A1, filed November 8, 2012, and issued September 15, 2010.

Wells, W. A. 2002. Is transdifferentiation in trouble? *Journal of Cell Biology* 157 (1): 15–18.

Wetter, O., H. Delbruck, and K. H. Linder. 1978. Surface markers on peripheral blood lymphocytes of patients with follicular lymphoma suggesting a clonal origin. *Klinische Wochenschrift* 56 (8): 415–419.

Whang, J., E. Frei III, J. H. Tjio, P. P. Carbone, and G. Brecher. 1963. The distribution of the Philadelphia chromosome in patients with chronic myelogenous leukemia. *Blood* 22: 664–673.

Wicha, M. S. 2014. Targeting self-renewal, an Achilles' heel of cancer stem cells. *Nature Medicine* 20 (1): 14–15.

Wicha, M. S., S. Liu, and G. Dontu. 2006. Cancer stem cells: An old idea—a paradigm shift. *Cancer Research* 66 (4): 1883–1890; discussion 1895–1896.

Wiestler, O. D., B. Haendler, and D. Mumberg, eds. 2007. *Cancer Stem Cells. Novel Concepts and Prospects for Tumor Therapy.* Berlin: SpringerVerlag.

Willingham, S. B., J. P. Volkmer, A. J. Gentles, D. Sahoo, P. Dalerba, S. S. Mitra, J. Wang, H. Contreras-Trujillo, R. Martin, J. D. Cohen, P. Lovelace, F. A. Scheeren, M. P. Chao, K. Weiskopf, C. Tang, A. K. Volkmer, T. J. Naik, T. A. Storm, A. R. Mosley, B. Edris, S. M. Schmid, C. K. Sun, M. S. Chua, O. Murillo, P. Rajendran, A. C. Cha, R. K. Chin, D. Kim, M. Adorno, T. Raveh, D. Tseng, S. Jaiswal, P. O. Enger, G. K. Steinberg, G. Li, S. K. So, R. Majeti, G. R. Harsh, M. van de Rijn, N. N. Teng, J. B. Sunwoo, A. A. Alizadeh, M. F. Clarke, and I. L. Weissman. 2012. The CD47-signal regulatory protein alpha (SIRPa) interaction is a therapeutic target for human solid tumors. *Proceedings of the National Academy of Sciences USA* 109 (17): 6662–6667.

Willis, R. A. 1958. *The Borderland of Embryology and Pathology.* London: Butterworth.

Wilmut, I., A. E. Schnieke, J. McWhir, A. J. Kind, and K. H. Campbell. 1997. Viable offspring derived from fetal and adult mammalian cells. *Nature* 385 (6619): 810–813.

Wilson, R. 1999. Realism, essence, and kind: Resuscitating species essentialism? In *Species: New Interdisciplinary Studies,* edited by R. Wilson. Cambridge, MA: MIT Press.

Wilson, R., M. J. Barker, and I. Brigandt. 2007. When traditional essentialism fails: Biological natural kinds. *Philosophical Topics* 35 (1&2): 189–215.

Witschi, E. 1948. Migration of the germ cells of human embryos from the yolk sac to the primitive gonadal folds. In *Contribution to Embryology,* edited by the Carnegie Institution. Washington, DC: Carnegie Institution.

Wolfort, B., and D. Sloane. 1938. Adamantinoma of the tibia. A report of two cases. *Journal of Bone and Joint Surgery* 20 (4): 1011–1018.

Wolkenhauer, O. 2001. Systems biology: The reincarnation of systems theory applied in biology? *Briefings in Bioinformatics* 2 (3): 258–270.

Wright, L. 1973. Functions. *Philosophical Review* 82 (2): 139–168.

Wurmser, A. E., and F. H. Gage. 2002. Stem cells: Cell fusion causes confusion. *Nature* 416 (6880): 485–487.

Xie, T., and A. C. Spradling. 1998. Decapentaplegic is essential for the maintenance and division of germline stem cells in the Drosophila ovary. *Cell* 94 (2): 251–260.

Yahata, T., K. Ando, T. Sato, H. Miyatake, Y. Nakamura, Y. Muguruma, S. Kato, and T. Hotta. 2003. A highly sensitive strategy for SCID-repopulating cell assay by direct injection of primitive human hematopoietic cells into NOD/SCID mice bone marrow. *Blood* 101 (8): 2905-2913.

Yamashita, Y. M. 2010. Cell adhesion in regulation of asymmetric stem cell division. *Current Opinion in Cell Biology* 22 (5): 605-610.

Yamashita, Y. M., D. L. Jones, and M. T. Fuller. 2003. Orientation of asymmetric stem cell division by the APC tumor suppressor and centrosome. *Science* 301 (5639): 1547-1550.

Yang, G., Y. Quan, W. Wang, Q. Fu, J. Wu, T. Mei, J. Li, Y. Tang, C. Luo, Q. Ouyang, S. Chen, L. Wu, T. K. Hei, and Y. Wang. 2012. Dynamic equilibrium between cancer stem cells and non-stem cancer cells in human SW620 and MCF-7 cancer cell populations. *British Journal of Cancer* 106 (9): 1512-1519.

Ye, J., D. Wu, P. Wu, Z. Chen, and J. Huang. 2014. The cancer stem cell niche: Cross talk between cancer stem cells and their microenvironment. *Tumor Biology* 35 (5): 3945-3951.

Yilmaz, O. H., R. Valdez, B. K. Theisen, W. Guo, D. O. Ferguson, H. Wu, and S. J. Morrison. 2006. Pten dependence distinguishes haematopoietic stem cells from leukaemia-initiating cells. *Nature* 441 (7092): 475-482.

Yoshihara, H., F. Arai, K. Hosokawa, T. Hagiwara, K. Takubo, Y. Nakamura, Y. Gomei, H. Iwasaki, S. Matsuoka, K. Miyamoto, H. Miyazaki, T. Takahashi, and T. Suda. 2007. Thrombopoietin/MPL signaling regulates hematopoietic stem cell quiescence and interaction with the osteoblastic niche. *Cell Stem Cell* 1 (6): 685-697.

Younes, S. A., B. Yassine-Diab, A. R. Dumont, M. R. Boulassel, Z. Grossman, J. P. Routy, and R. P. Sekaly. 2003. HIV-1 viremia prevents the establishment of interleukin 2-producing HIV-specific memory CD4+ T cells endowed with proliferative capacity. *Journal of Experimental Medicine* 198 (12): 1909-1922.

Yu, J., M. A. Vodyanik, K. Smuga-Otto, J. Antosiewicz-Bourget, J. L. Frane, S. Tian, J. Nie, G. A. Jonsdottir, V. Ruotti, R. Stewart, I. I. Slukvin, and J. A. Thomson. 2007. Induced pluripotent stem cell lines derived from human somatic cells. *Science* 318 (5858): 1917-1920.

Yu, J. S., ed. 2009. *Cancer Stem Cells: Methods and Protocols*. Dordrecht: Humana Press.

Zacks, S. I., and M. F. Sheff. 1982. Age-related impeded regeneration of mouse minced anterior tibial muscle. *Muscle Nerve* 5 (2): 152-161.

Zegarelli, E. V. 1944. Adamantoblastomas in the slye stock of mice. *American Journal of Pathology* 20 (1): 23-87.

Zhang, B., Y. W. Ho, Q. Huang, T. Maeda, A. Lin, S. U. Lee, A. Hair, T. L. Holyoake, C. Huettner, and R. Bhatia. 2012. Altered microenvironmental regulation of leukemic and normal stem cells in chronic myelogenous leukemia. *Cancer Cell* 21 (4): 577-592.

Zhang, H., H. Wu, J. Zheng, P. Yu, L. Xu, P. Jiang, J. Gao, H. Wang, and Y. Zhang. 2013. Transforming growth factor β1 signal is crucial for dedifferentiation of cancer cells to cancer stem cells in osteosarcoma. *Stem Cells (Dayton, Ohio)* 31 (3): 433–446.

Zhang, M., R. L. Atkinson, and J. M. Rosen. 2010. Selective targeting of radiation-resistant tumor-initiating cells. *Proceedings of the National Academy of Sciences USA* 107 (8): 3522–3527.

Zhang, S. O., and D. A. Weisblat. 2005. Applications of mRNA injections for analyzing cell lineage and asymmetric cell divisions during segmentation in the leech Helobdella robusta. *Development* 132 (9): 2103–2113.

Zhang, X., K. T. Ebata, B. Robaire, and M. C. Nagano. 2006. Aging of male germ line stem cells in mice. *Biology of Reproduction* 74 (1): 119–124.

Zhang, Y., G. Joe, E. Hexner, J. Zhu, and S. G. Emerson. 2005. Host-reactive CD8+ memory stem cells in graft-versus-host disease. *Nature Medicine* 11 (12): 1299–1305.

Zhou, B. B., H. Zhang, M. Damelin, K. G. Geles, J. C. Grindley, and P. B. Dirks. 2009a. Tumour-initiating cells: Challenges and opportunities for anticancer drug discovery. *Nature Reviews Drug Discovery* 8 (10): 806–823.

Zhou, H., S. Wu, J. Y. Joo, S. Zhu, D. W. Han, T. Lin, S. Trauger, G. Bien, S. Yao, Y. Zhu, G. Siuzdak, H. R. Schöler, L. Duan, and S. Ding. 2009b. Generation of induced pluripotent stem cells using recombinant proteins. *Cell Stem Cell* 4 (5): 381–384.

Zhou, W., and C. R. Freed. 2009. Adenoviral gene delivery can reprogram human fibroblasts to induced pluripotent stem cells. *Stem Cells* 27 (11): 2667–2674.

Zhou, Y., J. H. Morais-Cabral, A. Kaufman, and R. MacKinnon. 2001. Chemistry of ion coordination and hydration revealed by a K+ channel-Fab complex at 2.0 A resolution. *Nature* 414 (6859): 43–48.

Zhu, H., D. Wang, D. Liu, Z. Su, L. Zhang, F. Chen, Y. Zhou, Y. Wu, M. Yu, Z. Zhang, and G. Shao. 2013. Role of the Hypoxia-inducible factor-1 alpha induced autophagy in the conversion of non-stem pancreatic cancer cells into CD133+ pancreatic cancer stem-like cells. *Cancer Cell International* 13 (1): 119.

Zhu, W., G. M. Pao, A. Satoh, G. Cummings, J. R. Monaghan, T. T. Harkins, S. V. Bryant, S. Randal Voss, D. M. Gardiner, and T. Hunter. 2012. Activation of germline-specific genes is required for limb regeneration in the Mexican axolotl. *Developmental Biology* 370 (1): 42–51.

Zieker, D., S. Buhler, Z. Ustundag, I. Konigsrainer, S. Manncke, K. Bajaeifer, J. Vollmer, F. Fend, H. Northoff, A. Konigsrainer, and J. Glatzle. 2013. Induction of tumor stem cell differentiation—novel strategy to overcome therapy resistance in gastric cancer. *Langenbecks Archiv für Chirurgie* 398 (4): 603–608.

Zipori, D. 2004. The nature of stem cells: State rather than entity. *Nature Reviews Genetics* 5 (11): 873–878.

———. 2009. *Biology of Stem Cells and the Molecular Basis of the Stem State.* New York: Humana Press.

Zuk, P. A., M. Zhu, H. Mizuno, J. Huang, J. W. Futrell, A. J. Katz, P. Benhaim, H. P. Lorenz, and M. H. Hedrick. 2001. Multilineage cells from human adipose tissue: Implications for cell-based therapies. *Tissue Engineering* 7 (2): 211–228.

ACKNOWLEDGMENTS

This book is the outcome of over six years of research, funded by the Cancéropôle Île-de-France, the ARC Foundation, and the SIRIC-SOCRATE program. Over the course of this time, I have presented its contents to philosophers, biologists, and clinicians. Comments, questions, and discussions from these three communities have been critical in the constitution of my arguments. I had the chance to share in particularly fruitful research environments at the University Paris Ouest Nanterre, University Paris I, University Paris 6, and Hospital Gustave Roussy in France; the Max Planck Institute for the History of Science in Germany; and Arizona State University in the United States. I have discussed this work among philosophers and historians of science, including Bernadette Bensaude-Vincent, Sébastien Dutreuil, Melinda Fagan, Jean

Gayon, Thierry Hoquet, Jane Maienschein, Francesca Merlin, Michel Morange, Matteo Mossio, Antonine Nicoglou, Thomas Pradeu, and Karine Prévot; and among biologists and physicians, including Pierre Charbord, Hervé Chneiweiss, Charles Durand, Eve Gazave, Philippe Gorphe, Thierry Jaffredo, Leïla Perié, Françoise Pflumio, Eric Solary and his team, Michel Vervoort, and members of the French cancer stem cell network. All played a role in the development of the arguments found in this book through encouragement, discussion, critiques, and comments on previous versions of the work.

This book would not have been possible without Michael Fisher, Jane Maienschein, and Kate MacCord. I am grateful for their priceless help and support. Kate really did an amazing job that went far deeper than just editing the English. Federica Turriziani translated early versions of the first four chapters. It has been a pleasure to work with Harvard University Press, and I am grateful in particular to Lauren Esdaile and Thomas LeBien.

I am thankful to my family and my friends for their unconditional support, which is so important to me.

acute leukemia, 77–79, 99; acute lymphoid leukemia, 144; acute myeloid leukemia (AML), 24, 74, 81–82, 85, 102, 144, 147, 153, 186; acute promyelocytic leukemia (APL), 175–176; childhood acute lymphoblastic leukemia, 99
adamantinoma, 36, 56
antiangiogenic, 2, 143, 165, 176
antibody, 19, 22, 24, 83–87, 94–95, 147–148
antigen, 24, 84–85, 95, 147–148
Arai, Fumio, 128
Arechaga, Juan, 67–68
Askanazy, Max, 58, 65, 69
Astaldi, Giovanni, 76
Auffinger, Brenda, 164

Bapat, Sharmila, 99
barcoding, 108, 188
Barker, Matthew, 110–111
Barnes, David, 73
Beard, John, 58–60, 69
Becker, Andy, 81
bidirectional interconversion, 161–162
Billroth, Theodor, 53
binominal definition, 112–114
Bird, Alexander, 152
blast cell, 77–79, 87

blastocyst, 5, 57–58, 62–64, 66
blastomere, 5, 56–58, 60, 69
Blau, Helen, 135, 140, 181
blood, 35, 77–78, 171, 175–176; cancer, 4, 8, 23, 70; cell, 6, 8, 73, 75–76, 80–81, 85, 108, 133; disorder, 83, 165; stem cell, 4; vessel, 143–144
bone marrow transplantation, 4, 70, 79–82, 87
Bonnet, Dominique, 19–20, 94
Bonnet, Robert, 57, 60, 69
boundary object, 112–113
Boveri, Theodor, 75
Boyd, Richard, 110, 114
Brabletz, Thomas, 99
brain cancer, 13, 86, 143–144, 163–164. *See also* glioblastoma; glioma
Brawley, Crista, 132
breast cancer, 23, 86, 96, 103, 143, 148, 159–163, 176
Brigandt, Ingo, 110–111
Brues, Austin, 61–62, 66
Budde, Max, 65
Buhl, Ludwig, 53
Burian, Richard, 113

Cabarcas, Stéphanie, 143, 151
Cabral, Joaquim, 124
Cairns, John, 35, 98

Calabrese, Christopher, 143–154

Campos, Benito, 176

cancer-initiating cell, 37, 86, 92–101, 147, 158, 163

cancer-propagating cell, 92, 95–99, 101

cancer stem cell (CSC) targeting, 2–4, 36, 41–43, 178, 180, 182–183, 187; difficulties, 140, 145–147; drugs, 20, 22–26, 174–176; efficiency in function of the identity of stemness, 149, 151–152, 155–157, 164–166; identifying and targeting, 17. *See also* therapy

cancer stem cell (CSC) theory, 3, 8, 98, 145, 179–180, 181; CSC theory ambiguities, 88, 92, 180; CSC theory and stemness identity, 149, 152, 171–172, 182; CSC theory and therapy, 20–21, 43, 135, 148, 164, 178; CSC theory origin, 47–48, 63, 69–70, 87, 91, 104, 145, 175; CSC theory prediction, 147, 151; CSC theory proponents, 146, 175; CSC theory structure and content, 27, 29, 31–44, 94, 100, 156–157, 185

cancer stem-like cell, 92, 102, 161

cancer stemloid cell, 92, 102. *See also* cancer stem-like cell

carcinogenesis: cancer stem cells as passengers of carcinogenesis, 163; CSC model of, 27, 29–36, 39–40, 43–44; the evolution theory of, 98; Pierce view of, 66, 68; precancerous, 146

categorical property: classification, 8–9, 155–156, 180–183, 186–188; consequence for therapies, 145–147, 149–151, 177; vs. dispositional property, 140–142; vs. systemic property, 172–173

category, 7–8, 142; of cells, 29, 55, 63, 88, 132; of stem cells, 109, 111, 113, 134–135, 180

causal bases of stemness, 144, 149, 151

CD47, 148–151

cell culture, 5, 63–64, 80–82, 85, 108, 123, 130, 160

cell fate, 120, 125, 132, 134, 144

cell lineage: cancer, 28, 34, 39, 98; stem cells, 88, 107, 166; tracing, 72, 80, 188

cell of origin in cancer, 48, 96

cell sorting, 80, 84, 86, 95. *See also* FACS

cell type: cancer, 41, 59; cancer initiation, 33, 51, 63; differentiated cells, 5, 36, 59, 73, 82, 107–111, 124; hematopoietic, 73, 76, 80, 82, 108; stem cells, 4–5, 60, 74, 91, 128, 135, 181, 186; system biology, 170; tracing, 188; transdifferentiation between cell types, 133–134

Ceteci, Fatih, 163

Chaffer, Christine, 177

Chiquoine, Duncan, 59–60

Chneiweiss, Hervé, 163

chronic lymphocytic leukemia (CLL), 74

chronic myeloid leukemia (CML), 43, 74, 153, 177

Clarke, Michael, 22, 101

Clarkson, Bayard, 17, 79

classification, 8–9, 50, 65, 109, 135

Clevers, Hans, 19

clonal evolution, 15, 34–35, 37–38, 71, 98–99, 145

cloning, 120–122, 124–125, 134–135, 141

clonogenicity, 81–82, 160; assay, 82–83; cancer cells, 32–35, 40, 43; definition, 34

Cohnheim, Julius, 48, 50, 53–55, 57–58, 60, 65, 69

colorectal cancer, 22, 24, 41, 99, 160, 165, 177

Colvin, Gerald, 150

Conboy, Irina, 130

connective tissue theory, 48, 51–53, 69

Crist, Eileen, 112

Cronkite, Eugene, 77
Cummins, Robert, 171–172

Dalprà, Leda, 99
Dantschakoff, Wera, 75
Darwin, Charles, 98, 110
Dazard, Jean-Eudes, 116
dedifferentiation, 135, 185–188; in
    cancer, 160, 163–166, 174–175,
    177–178; cellular plasticity, 134, 141;
    cloning, 121–122, 141; iPS cells,
    122–124, 141; niche-induced,
    127–128, 132, 141; Pierce, 66;
    probability of, 159–160, 174–175,
    177–178, 185, 187; regeneration,
    125–127; STAP cells, 124–125, 141
definition, 7; Aristotelian, 110;
    binominal, 112–114; of cancer stem
    cells, 27–28, 29, 101, 104, 172, 180;
    by clusters, 110, 114; of function,
    170–172; generic vs. specific,
    112–114; homeostatic property
    cluster, 110–114; molecular
    definition of stemness, 115; of stem
    cells, 6, 9, 107–109, 180; of stemness,
    29, 181; stringent vs. fuzzy, 112
delocalized embryonic rest theory, 48,
    51–56, 69
de Thé, Hugues, 176
development: of cancers or tumors, 76,
    92, 161, 163, 172, 179; of chronic
    myeloid leukemia, 43; context,
    185–186; CSC theory of the develop-
    ment of cancer, 27–28, 31, 35, 43, 86,
    98, 147, 172; of the embryo, 5, 51,
    53–55, 57, 116, 122, 133; embryonic
    theories of cancer development,
    48–49, 52–54; fate, 144; irreversibity,
    120; of mice, 64, 66; pathological, 59;
    periods, 4–5, 62–63; potential, 69, 91,
    125, 128, 130; stem cells, 55; of
    teratoma/teratocarcinoma, 59,
    62–63, 65; of tissues, 68, 103
Dezawa, Mari, 123

Dick, John, 39, 83, 85, 92, 94–95, 99
differentiation: blastomeres, 57;
    blockage, 176; in cancer, 51, 53,
    65–67; cancer stem cells, 3, 28–30,
    33, 55; commitment, 74, 76, 102;
    definition, 107–109; gene expres-
    sion, 120, 122; hematopoietic cells,
    81, 85, 87; hierarchy, 82, 120, 127,
    132, 134–135, 188; irreversiblity, 67,
    120–122, 131; neoplastic stem cells,
    68; pluripotent stem cells, 60;
    regulation by the niche, 128–129,
    131, 144, 165; reversibility, 120,
    124, 127, 132, 135; self-maintaining
    system, 77; states, 175; stem cells, 5,
    63, 88; stemness, 6–7, 38–39, 115,
    130, 140, 150, 180–181, 185. See also
    differentiation therapy
differentiation therapy, 3, 175–178,
    180, 182–184
Dirks, Peter, 13, 15
Discher, Dennis, 130
dispositional property, 54; in cancer
    stem cells, 144–145; classification,
    8–9, 155–156, 177, 180–183,
    186–188; consequences for therapy,
    149–153, vs. categorical, 140, 142;
    vs. function, 170–174; vs. relational
    property, 168
Dixon, Frank, 65, 67

Eaves, Connie, 19
Ehrlich, Paul, 75
Eisenberg, Carol and Leonard, 134
embryo, 4, 51, 57–59, 62–64, 73, 75,
    109, 111, 122
embryoid body, 62, 64, 67
embryoma, 57–59
embryonic/embryonal carcinoma (EC)
    cell, 5, 63–64, 66–67
embryonic rest theory, 47–48, 51–57,
    60, 65, 69
embryonic stem (ES) cell, 5, 8, 55–56,
    60, 63–64, 115–116, 122, 134, 163

entity view, 139–141, 181, 187

Enver, Tariq, 96

epithelial to mesenchymal transition (EMT), 159–160, 166, 176, 186

essentialism, 109–110; essential property, 119, 132, 140–142, 155, 181

Essers, Marieke, 152–154

Evans, Martin, 64

evolution, 110, 112, 114, 171. *See also* clonal evolution

extrinsic property, 140, 156–157, 164, 166–168, 177–178, 181

extrinsic relational property, 166–169, 178, 181, 183, 186–188

extrinsic stimulus/factor, 140, 142, 155, 166, 174, 186–187

FACS, 80, 84–87, 160–161

Fagan, Melinda, 86, 107, 173

Farrar, William, 143

feeder-sleeper stem cell model, 78

Fekete, Elizabeth, 59, 61–62, 65

Ferrigno, Mary Ann, 59, 61–62, 65

Fialkow, Philip, 74–76

Fisher, Virginia, 77–78

Fojo, Tito, 165

Ford, Charles, 80

Fouchet, Pierre, 132

Fried, Jerrold, 79

Frisen, Jonas, 135, 140

function: of cancer stem cells, 29, 55; of cancer-initiating cells, 92; non-dispositional systemic, 170–174; of stem cells, 131; stemness as a function, 140, 156, 166, 170, 181; systemic, 157, 181

gastric cancer, 22, 177

Gavosto, Felice, 77–79

genetic or genomic instability, 34, 73, 98

Gerber, Jonathan, 99

germ layer, 5, 51–52, 56–57, 120

Ghiaur, Gabriel, 99

Gilbertson, Richard, 19

Gilliland, Gary, 102

Glatzle, Jörg, 177

glioblastoma, 24, 99, 164

glioma, 144, 163–164, 176

Golan-Mashiach, Michal, 116

Grácio, Filipe, 124

Grady, Christine, 165

Greaves, Mel, 96–99

Gurdon, John, 122

Haeckel, Ernst, 4, 75

Haecker, Valetin, 75

Hanna, Jacob, 124

Hannover, Adolph, 51

hematopoietic stem cell: competition with leukemic stem cells, 144; concept of, 87; definition, 113–115; disposition, 186, 189; feeder-sleeper model, 78; history, 6, 8; identification, 81–83, 85, 131; markers, 116, 119, 148, 151; niche, 128, 153–154; at the origin of blood cells, 73; origin of cancer stem cell, 39; plasticity, 133; preleukemic stem cells, 96

Herlyn, Meenhard, 100

His, Wilhelm, 53

Hochedlinger, Konrad, 122, 124

homeostatic property cluster (HPC), 109–114

Huang, Sui, 170

Hull, David, 110, 114

Humberstone, Lloyd, 167

identity, 134, 167; cancer stem cell, 2, 53, 92, 172–173, 180, 182; categorical, 145–146, 155; categorization of stem cell identities, 139, 157–158, 182, 185, 187; dispositional, 145, 155; niche, 135, 181; objects, 7; stem cell and the niche, 128, 131–132; stem cell and oncology, 4, 8–9; stem cell and regeneration, 125; stem cell and system biology, 169; stemness, 8, 105, 115, 156, 180, 185–187

Illmensee, Karl, 64, 66

induced pluripotent stem (iPS) cell, 5, 115, 122–125, 127, 134–135, 141, 163

interconversion, 79, 161–162

intrinsic property: cancer stem cell, 33, 63; categorical vs. dispositional property, 140–142; causal bases, 144, 149, 151; cell, 134; disposition, 173; embryonic cell, 53; leukemic stem cells, 7, 73; regulated or controlled by the niche, 130–132; stemness, 87, 119, 155–157, 180–181, 187; stochastic, 174, 178; vs. extrinsic property, 166–168

Izppisua Belmonte, Juan Carlos, 127

Jackson, Elizabeth, 61–62, 65

Jaenisch, Rudolf, 122

Jones, Richard, 99

Jopling, Chris, 127

Junier, Marie-Pierre, 163

Kaufman, Matthew, 64

Killmann, Sven-Aage, 77–78

Klebs, Edwin, 53

Kleinsmith, Lewis, 67

Köhler, George, 85

Lagasse, Eric, 99

Lajtha, Laszlo, 75

Lambert, Jean-François, 150

Lander, Arthur, 166

Lander, Eric, 160–162, 176

Langhans, Theodor, 53

Lee, John, 100

Lemischka, Ihor, 119

leukemia: history, 70–74, 76, 78, 81–84, 86–87; mouse model; 92–94, 102–103; survival rate; 2; treatment, 148, 154. *See also* acute leukemia; chronic lymphocytic leukemia;

chronic myeloid leukemia; lymphoblastic leukemia; myeloid leukemia

leukemic stem cell, 8, 24, 102–103, 144, 176; heterogeneity, 114; history, 70, 73–74, 76–80, 87

Lewis, David, 167

Lim, Bing, 119

Lin, Chia-Ying, 176

Lindeman, Geoffrey, 101

Little, Clarence Cook, 61

Löwy, Ilana, 112

lymphoblastic leukemia, 99

Magee, Jeffrey, 162

Malanchi, Ilaria, 144, 154

Marchand, Felix, 57, 60, 69

Marjanovic, Nemanja, 177

Martin, Gail, 64

Mathews, Lesley, 143

Matunis, Erika, 132

Mauer, Alvin, 77

Mauri, Carlo, 76

Maximow, Alexander, 75

McCulloch, Ernest, 75, 81–82, 131

Medema, Jan Paul, 160, 162, 165

Melton, Douglas, 119

metaphysics, 140, 142–143, 145, 151, 155, 166–167, 169, 186

metaplasia, 52, 132, 134

metastasis, 36, 98–100, 144–145, 147, 152, 163; non-metastatic cancer cell, 32–35, 41, 43

microenvironment, 111, 119, 127–132, 142–144, 160, 163, 166. *See also* niche

Mikkers, Harald, 135, 140

Milstein, César, 85

Mintz, Beatrice, 60, 64, 66

Mitalipov, Shoukhrat, 122

model: analogical model, 152; CSC model by Tannishtha Reya et al., 21–22; CSC theory's models, 27–40, 42–44, 101, 156–157, 185; elite vs. stochastic iPS models, 123–124; feeder-sleeper stem cell model, 78;

model (continued)
hierarchical model of differentiation, 82, 86, 120, 133–134; hierarchical vs. dynamic and bidirectional interconversions models of cancer stem cells, 159–163, 165, 176; initiation-propagation model, 103; for metastatic cancer stem cells, 99; Pierce's model of cancer development, 68; Stevens's model of cancer development, 63; stochastic model of clonal evolution, 98
model organism, 4, 61, 94–95, 128; mouse model, 61, 80, 83–84, 102–103, 154
molecular signature of stemness, 24, 106, 108, 115–119, 135, 146–151, 155
monoclonal antibody, 22, 83–87, 148
Moore, Malcolm, 81
Moore, Robert, 65
Morrison, Sean, 22, 32, 94, 154, 162
Müller, Johannes, 50
Mulloy, James, 144
multiple myeloma, 23, 147–148
multipotent stem cell, 6, 75, 84, 107, 114–115, 134
myelodysplastic syndrome, 147
myeloid leukemia, 102

nanovector, 146, 165
natural kind, 7, 9, 109–111, 119–121, 132–135, 140–142, 155, 180
Neel, Ben, 25
niche: cancer stem cell, 143–144, 150–151, 160; controlling stem cells (weak interpretation of the niche hypothesis), 128–130, 135, 141, 155–157; hematopoietic stem cell, 151, 153, 189; hypoxic niche, 38; inducing stemness (strong interpretation of the niche hypothesis), 128, 132, 135, 141, 156, 178, 185; niche targeting, 3; philosophical analysis, 152–153, 168–169; Schofield, 127,

159; stem cell definition, 111, 113, 134, 181, 183, 187; targeting, 3, 155, 165–167, 174–175, 178, 180, 183, 187–188
Nixon, Richard, 1
Nowell, Peter, 35, 73, 98

Obokata, Haruko, 124–125
Ogura, Atsuo, 122
osteosarcoma, 164, 177
ovarian cancer, 24, 148

pancreatic cancer, 2, 23, 164
Pandolfi, Pier Paolo, 153
Pappenheim, Arthur, 75
Peyron, Albert, 58–59
Pierce, Gordon Barry, 48, 56, 60, 64–68, 175
Piskounova, Elena, 162
plasticity/plastic, 121, 125–126, 132, 134–135, 141, 163, 186
pluripotency, 1, 64, 107, 111, 125, 127
pluripotent stem cell: in cancer, 48, 57, 59, 61, 74; cultured, 63; definition, 5, 107; gene expression, 127; induced, 122–124, 134; at the origin of cancer, 68; stem cell classification and definition, 74–76, 109, 114–115
Poss, Kenneth, 125
de Poulton Nicholson, Gilbert William, 58–59
Pradeu, Thomas, 112
prostate cancer, 2
Puisieux, Alain, 160

Quesenberry, Peter, 119, 150
quiescence, 38, 113, 120, 128–130, 150–151, 153, 155, 189

Rak, Janusz, 158–159
Rapp, Ulf, 163
regeneration, 116, 125–127, 133–135, 141
relapse: from cancer stem cells, 21, 151, 157, 178, 183; CSC model of,

27–28, 32, 35–37, 39–43, 179; in
various cancers, 163–164, 177
relational property, 119; classification,
8–9, 172, 180, 183, 186–188;
relational extrinsic, 166–169, 178,
181, 183, 186–188; vs. systemic
property, 156, 178, 181
Remak, Robert, 48–56, 69
remission, 37, 40–43, 153, 163
Reya, Tannishtha, 21, 28, 32, 35–37,
39, 93
Risbridger, Gail, 19

Schleiden, Matthias, 48–49
Schofield, Ray, 127, 131–132, 159
Schreck, Ralf, 163
Schwann, Theodor, 49–50, 52
self-renewal: blockage, 22, 165, 176;
cancer stem cells, 28, 30, 33,
102–103, 143; definition, 107–109;
gene expression, 120, 122, 146, 150,
165; hematopoietic and leukemic
stem cells, 81, 87, 113, 131, 189;
heterogeneity, 7; neoplastic stem
cells, 68; regulation by the niche,
128–129, 144; stem cells, 63, 88,
109–110, 130, 140; stemness, 3, 29,
39, 115, 180–181, 185
Sell, Stewart, 175
Shelton, Emma, 80
Song, Xiao-qing, 129
squamous cell carcinoma, 144, 154
Stadtfeld, Matthias, 124
STAP cells, 124–125, 141
state view, 140, 156, 166, 181, 187
stemness: acquisition, 7, 105–106, 141,
157–160, 163, 174, 186; as a
categorical property, 140–141,
145–149, 155–156, 177; as a
dispositional property, 141–142,
144–145, 149–153, 155–156, 177, 181;
as a multicellular or unicellular
property, 87, 159; as a non-
dispositional systemic function,

170–174; as a relational extrinsic
property, 156–157, 166–169; as a
systemic property, 157, 166,
169–170, 174–175, 178, 181; as a
transient state, 135, 158–159; as an
extrinsic property, 140, 156–157,
164, 168, 177–178, 181, 186; as
an intrinsic property, 156, 167,
180–181; as the common feature of
stem cells, 6–7; cancer stem cells, 38,
43–44, 100, 179; controlled by the
niche, 128–131, 135, 141, 152,
155–156, 181; definition, 29,
107–108, 180, 185; identity, 4, 7–9,
139, 158, 180–189; induced by the
niche, 128, 131–132, 135, 141, 160,
166–167, 178, 181; molecular
signature, 24, 106, 108, 115–120,
135, 146–151, 155; property, 3, 86;
resistance to therapy, 38–39, 164,
183; retrospective identification,
108; stem cells, 68, 185
Stevens, Leroy, 48, 56, 60–67, 69
Strasser, Andreas, 96
Strathern, Marilyn, 168
Suda, Toshio, 128
system biology, 169–170, 177
systemic property, 8–9, 157, 166,
169–175, 178, 180–183, 186–187

Takahashi, Kazutoshi, 123
Tauber, Alfred, 112
teratoma/teratocarcinoma, 5, 8, 48,
56–68
theory, 147, 149; blastema theory, 48,
56–58, 60, 69; cell theory, 48–50, 52,
69, 180; clonal evolution theory, 15,
98; connective tissue theory, 48,
52–53, 69; delocalized embryonic
rest theory, 48, 51–53, 69; embryonic
rest theory, 47, 51, 55–56, 60; germ
cell theory, 48, 56–58, 60, 69; germ
layer theory, 51–52, 120; homeostatic property cluster (HPC) theory,

theory *(continued)*
110–111; monophyletic theory of hematopoiesis, 73; niche theory, 131; Pierce theory of cancer development, 66–68; somatic mutation theory, 66; superabundant embryonic rest theory, 48, 53–54, 69; theory of disease, 49; theory of the development of teratoma/teratocarcinoma, 65; theory of humors, 49–50; trophoblastic theory, 58, 60. *See also* cancer stem cell (CSC) theory

therapy, 8, 24–26, 36, 152–153; anti-angiogenic, 2; chemotherapy, 3, 36–37, 164–165; combined therapies, 164, 183, cost of therapies, 164–165, 183; CSC therapeutic model, 41–43; differentiation therapy, 3, 175–178, 180, 182–184; gene therapy, 83; immunotherapy, 2, 15, 23–24, 163; niche-targeting, 155, 165–167; radiotherapy, 3, 37; stem cell therapy, 4, 81, 133, 188–189; targeting therapy, 2; therapeutic hope, 20; therapeutic strategy, 2–3, 21–22, 28; therapy and stemness identity, 139, 145, 149, 155, 158, 174, 178, 180, 182–185, 187–188; therapy resistance or escape, 36–41, 43, 151, 164, 179; traditional therapies, 21, 41–42. *See also* cancer stem cell (CSC) targeting

Thiersch, Carl, 53
Thomson, James, 64
Tidor, Bruce, 124
Till, James, 81–82, 131
totipotent stem cell, 5, 107, 125
transdifferentiation, 121, 132–135
Trumpp, Andreas, 25, 152

Tu, Shi-Ming, 101
tumorigenicity, 3, 63, 145, 149, 152, 155, 159–160
tumor-initiating cell. *See* cancer-initiating cell
Tüting, Thomas, 163

Varet, Bruno, 102
Vermeulen, Louis, 165
Vervoort, Michel, 115
Vescovi, Luigi Angelo, 99
Virchow, Rudolf, 48–53, 55, 69
Visvader, Jane, 19, 96–98, 101

Waddington, Conrad, 169
Wakayama, Sayaka, 122
Wang, Yugang, 162
Weber, Carl Otto, 53
Weinberg, Robert, 159–160, 162, 166, 177
Weissenberg, Richard, 75
Weissman, Irving, 32, 84–85, 121, 147–150
Wicha, Max, 103, 143
Willis, Rupert Allan, 65
Wilms, Max, 57, 60, 69
Wilson, Darcy, 83
Wilson, Robert, 110–111
Witshi, Emil, 59–60

Xie, Ting, 129

Yamanaka, Shinya, 122–123

Zhang, Bin, 144
Zhou, Da, 162
Zhu, Wei, 127
Zipori, Dov, 109, 116, 124, 134, 135, 140, 181